TECHNOLOGY and the TRANSFORMATION of WHITE-COLLAR WORK

TECHNOLOGY and the TRANSFORMATION of WHITE-COLLAR WORK

edited by
Robert E. Kraut
Bell Communications Research
Morristown, New Jersey

LEA
1987
LAWRENCE ERLBAUM ASSOCIATES, PUBLISHERS
Hillsdale, New Jersey London

Lawrence Erlbaum Associates, Inc., Publishers
365 Broadway
Hillsdale, New Jersey 07642

Library of Congress Cataloging in Publication Data

Technology and the transformation of white-collar work.

Presented at a conference sponsored by Bell Communications Research, June, 1984.
Bibliography: p.
Includes indexes.
1. White collar workers--United States--Effect of technological innovations on--Congresses. 2. Office practice--Automation--Congresses. I. Kraut, Robert E. II. Bell Communications Research, Inc.
HD8039.M4U577 1987 331.7'92'0973 86-29039
ISBN 0-89859-633-5

Printed in the United States of America
10 9 8 7 6 5 4 3 2 1

Contents

Preface

It is now a commonplace to proclaim that rapid changes in communication and computer technology will radically transform the way in which white-collar work in the United States is done. For many, the implication is that although some may be buoyed up and others towed under by the tide of technology, the tide is inexorable. Yet the computer and telecommunications technologies are flexible and demand that we make choices about their use. These technologies have the potential to deaden and routinize challenging work or to free people from work that is now deadening and routine. They allow people to work at places and times that suit them, but they also allow employers to exploit workers or to transport jobs to regions with cheap and docile labor supplies. They have the potential to increase productivity or to disorient offices and to squander resources.

Can computers and telecommunications be introduced into the white-collar workplace in a manner that both increases the quality and quantity of the goods and services that organizations produce and also increases the quality of working life for the people who produce these goods and services? The present volume brings together leading scientists and practictioners to address these choices. Most of the chapters in this volume were commissioned for a conference on technology and the transformation of white-collar work that was sponsored by Bell Communications Research in June 1984. The authors are social scientists from America and Europe, from academe and from industry, interested in exploring the future and in understanding the past. They share a common desire to understand the ways in which white-collar technology can be introduced into the workplace to exploit its usefulness without exploiting its users.

The chapters in this volume examine the history, both ancient (the 1910s through the 1930s) and modern (the 1960s and on), of the introduction of white-collar technology. They trace management practices, advertising myths, social effects, and accuracy of previous predictions about social effects. This volume also surveys the contemporary scene, assessing the tasks for which the new technologies are best suited and the techniques that have been used successfully and unsuccessfully in introducing computers into the workplace, and prescribes methods that are likely to lead to the humane use of effective business tools. The impact of technology on women workers is the theme of several chapters, because so much of women's work is clerical work and because clerical work has been so heavily affected by automation.

This is a book for scholars and adopters of computer and telecommunication technologies, in universities, on office automation committees and in human resource departments. It is organized in two sections: Social impact of white-collar technology and Implementing technology in the office. The first section focuses on the impact of white-collar technology on employees, their numbers, compensation, and quality of employment. The authors adopt a contextual perspective that treats technology as embedded in managerial, organizational, and social contexts. The second section addresses the dynamics of designing white-collar technology and introducing it into the office. The authors argue that by understanding both the nature of office tasks and the processes by which technology is adopted, managers can successfully bring it to the office so that it is useful, used, and satisfying.

Robert E. Kraut

I SOCIAL IMPACT OF WHITE-COLLAR TECHNOLOGY

1 Social Issues and White-Collar Technology: An Overview

Robert E. Kraut
Bell Communications Research

ABSTRACT

We are in the midst of an information revolution. Decreasing prices and increasing capabilities of computer and telecommunications systems will push these technologies into ever expanding use. This chapter outlines issues necessary to understand the ways these new technologies have been and can be used to change the nature of white-collar work. What will happen to the numbers of white-collar jobs? What will happen to their quality? And can we predict these changes at all? The rest of the book attempts answers.

THE REVOLUTION IN WHITE-COLLAR WORK

The history of work has been, in part, the history of gradual changes in work technologies. Yet we term some of these changes revolutionary because their cumulative effects are so large. The industrial and agricultural revolutions took decades, but had the cumulative effect of transforming how manufactured goods and food are produced and how factories and farms are organized. They also had major indirect influences on many other aspects of society, including where people live, how they spend their time, and how their families are organized.

We are now in the midst of an information revolution. Both the visionaries and the doomsayers proclaim it. Certainly as Baker (1973) and Iacono and Kling (this volume) have documented, office work has had its share of technological innovations over the decades. But primarily because of declining costs, computer and telecommunications technologies are now influencing a wider range of tasks and a wider range of workers than did previous innovations. As a result they have the potential to make fundamental alterations in all white-collar work.

White-collar technologies include the computer and telecommunications hardware and software useful for the collection, processing, and transmission of voice, text, and pictorial information. But they also include the social arrangements in which this hardware and software are used. The hardware costs have dropped exponentially and their processing and transmission power per constant dollar has risen astonishingly (e.g., Caporeal, 1985). For example, in the 1960s, to computer typeset the million characters in *Moby Dick* would have required a mainframe computer, costing hundreds of thousands of dollars; today one can do this task on a personal computer, costing hundreds of times less. Similarly, because of advances in communications technologies—satellite, microwave, and fiber optic transmission—the costs and capabilities of communications systems also have been changing rapidly. The slow transmission facilities of the 1960s would have taken more than 9 hours to send the contents of Melville's novel over telephone lines; with the fiber optics of the 1990s, transmission could take less than 1 second.

By some estimates, 1980 was a cross-over point in which an hour of computer time became cheaper than an hour of a person's time (D. Smith, 1984). As a result, computers have been and will be used in ever expanding applications, both replacing people and aiding them. As electronic information becomes cheaper to collect, produce, transform and distribute, more of us will be both information workers (Porat, 1977)—the people who do the collection, production, transformation, and distribution—and information consumers. Writers reporting the latest fashion trends, receptionists announcing a telephone number, architects designing schoolhouses, insurance adjusters calculating the damage to our cars, clerks recording expense vouchers, teachers giving assignments, merchants calculating their income taxes, and spies assessing enemies' troop strength will all be using computers and telecommunications technologies to get their jobs done. Their fingers, eyes, and minds will be connected to silicon chips with wires of copper and glass.

There are of course some ways in which the current workplace revolution differs from earlier ones. Perhaps most important is the type of human labor for which machines are now substituting. In both the industrial and the agricultural revolution and in earlier changes in the white-collar workplace, machines substituted for and augmented human and animal physical labor. Machines worked the looms, turned the soil, and made the copies. Now the new microcomputer technology is substituting for and augmenting mental labor as well. Computers have the capability to recognize and categorize objects, for example to "read" the handwritten zip codes on letters. They have the ability to perform complex symbolic computations on and manipulations of the information provided to them. They can perform prodigious feats of arithmetic, auditing income tax returns, or tabulating the death rate associated with cigarette smoking in seconds. Or they can format a book, including proportional spacing, indexing, and creating a table of contents, and provide a stylistic analysis of text, pointing to misspelled words, split infinitives, dangled participles, or mangled language (e.g., Cherry & Vesterman, 1979). Or they can make decisions, grading a student's computer program (Soloway, Rubin, Woolf, & Bonar, 1983), diagnosing an illness (Shortliffe, 1976), or granting a loan. These tasks, once reserved for human recognition and reason, for postal clerks, statisticians, accountants, copyeditors, professors, doctors, and loan officers, can now be done by machine.

What is more, compared to earlier white-collar technologies, the current information technologies are more flexible. Not only can computers categorize, compute, analyze, and decide, but the same computer can do it all. Swapping different software into the same box causes that box to become a new tool, capable of performing new tasks. Furthermore, this flexibility is available to the actual user of the computer. With high level and often application specific programming languages, they can modify the way the machine works for their own uses.

The flexibility in the tasks that microelectronics can perform is paralleled by the flexibility in the manner in which they can be used and, as a result, in the social effects they are likely to have. Perhaps the machinery used to automate heavy industry severely constrained the way in which factory work could be organized and required that assembly-line workers perform small, monotonous tasks repetitively, under tight time pressure (but see Braverman, 1974; Walton, 1979 for alternate views). But white-collar technology, per se, doesn't force the same constraints. Word processing equipment can be organized with transcription secretaries in typing pools or with professionals composing on a computer without secretaries. Microelectronics allow easy monitoring and surveillance of workers by management, but don't require it. Using browsing programs, employers can search an employee's electronic files more easily than searching paper files, but using encryption and security programs, employees can safeguard their files. Computers can automatically determine the pacing of work, distributing customers to order entry clerks or to operators as soon as they have completed a transaction (Maddox, 1977); but one could as easily arrange the distribution so that the worker called for a new customer on her own schedule.

In all of these choices about the structure of work, the technology, if not neutral, is at least flexible. The bulk of the choices about how work is organized reflect social decisions, not technological mandates. This is a theme we return to repeatedly in this book. Bikson, Taylor, and Bjørn-Andersen and Kjærgaard in particular, in the chapters that follow, explore in depth the alternate ways of introducing white-collar technology.

The third way in which the microelectronics revolution differs from earlier changes in workplace technology is the extent to which it affects women as opposed to men. Although the new white-collar technology over the long term has the potential to alter the nature of managerial, professional, and clerical work (see Chamot, this volume), to date the influence has been strongest on clerical work. And clerical work is women's work. According to 1982 U.S. Labor Department figures, 85% of clerical workers were women, and over 95% of some clerical job categories, such as secretary, receptionist, typist, and telecommunications operator, were held by women (U.S. Bureau of the Census, 1983). These clerical jobs have been especially enmeshed in the microelectronics and telecommunications revolution.

In addition, as Iacono and Kling (this volume) show, clerical women tend to be at the bottom of organizational power hierarchies, and at least in the United States generally do not have union or other organized representation. As a result, women both individually and collectively have less ability to influence the use of these new technologies than do other affected parties.

Finally, because women often have child-care responsibilities and because they often are the lower paid worker in a two-job family, they have less freedom to individually adapt to the new technologies. Taking a training course or moving to a new city to follow the jobs may be more difficult for women than for men.

THE SAD HISTORY OF PREDICTIONS

A tenet for the authors in this book is the importance of studying social aspects of technology. The scope of the changes that the information revolution promises to bring to white-collar work should be sufficient to justify studying this topic. We need this study to make informed decisions to accomplish our business and social purposes, to manage the introduction of the technology within firms and throughout society, and to monitor the ripple of changes associated with the technology, evaluating unforeseen effects. Only in that way will we be able to exert conscious choice about how to use the technology and not be controlled by it.

Social Impact of the Telephone

Examining the history of predictions about the telephone and its actual impact is humbling and instructive. It warns us about the scope of the changes that might occur when adopting new white-collar technologies and about our abilities to predict them. Early in the history of the telephone, its creators and implementers failed to predict the extent to which it would be used and the uses to which it would be put. And certainly, in the early days, decision makers had little inkling of the profound changes in the social fabric to which the telephone contributed. We can draw a lesson from this history. We should cast our nets widely in examining social aspects of the new technology (cf. Lepper, 1985) and collect data as the changes are occurring. In this way we can evaluate the manner in which technology is being used, and the changes in productivity, employment levels, and employment quality associated with its use, in order to make informed decisions.

Bell had his historic conversation with Watson on March 10, 1876. By March 1878 Bell had envisioned a national switched network of business and residential telephones for interpersonal communication, through which one telephone subscriber could talk to any other in the country, even though this vision was a technological impossibility at the time (Pool et al., 1977). Bell himself invented the telephone while devising improvements to telegraphy. Although he conceived of the telephone as an instrument for talking at a distance, he also thought that it would be used as a broadcast device, similar to current cable TV.

In contrast to Bell's semi-prescience, many other technologists and capitalists of the time were less sure of the telephone's uses or success. During the late fall or winter of 1876–1877, William Orton, president of the Western Union Telegraph Company, turned down an offer to buy all rights to the telephone patent with the words, "What use could this company make of an electronic toy?" (Aronson, 1977, p. 16). Officials of the British Post Office Department also dismissed an offer to purchase Bell's English patent. Indeed, in 1879 Sir William Preece, the chief engineer of the British Post Office, testified to a special committee of the House of Commons that the telephone had little future in Britain:

> I fancy the descriptions we get of its use in America are a little exaggerated, though there are conditions in America which necessitate the use of such instruments more than here. Here we have a superabundance of messengers, errand boys and things of that kind. . . . Few have worked at the telephone much more than I have. I have one in my office, but more for show. If I want to send a message—I use a sounder or employ a boy to take it (in de Sola Pool, 1977, p. 128).

Social commentators trying to understand the impact that the telephone would have on social life had a less successful record than did Bell, Vail, the first head of AT&T, and other business entrepreneurs predicting its technological and business future. For example, some believed that the telephone would promote world peace by encouraging communication among nations, that it would reduce the volume of crime by allowing citizens to contact the police easily, that it would eliminate the need for travel, that it would democratize society and hierarchical organization, that it will lead to centralization of command at higher echelons among executive, diplomats, and military personnel, or that it would lead to increased job mobility (Pool, 1983). In part, this contrast in accuracy occurred because Bell, Vail, and the other capitalists and entrepeneurs were activists and thus in some position to shape the future, while commentators can only predict. But, as de Sola Pool and his students (Pool et al., 1977) amply demonstrate, social commentators and historians had relatively little to say about the telephone, compared to other technological inventions, such as the railroad, automobile, radio, or television.

Second, the commentators failed to recognize that the telephone was a permissive technology; the compression of space that the telephone allowed had diverse effects (Gottmann, 1977). The same telephone that relieved the social isolation of the farmer and induced people to stay in rural areas, also relieved the social isolation of the urban elderly who feared to walk the streets and induced them to stay in cities. The same telephone that supported the breakdown of traditional extended families also kept families in touch who would otherwise be separated by geographic mobility. And the same telephone that helped spawn the suburbanization of American cities, also fostered the growth of congested urban areas. In particular, the telephone allowed companies to move out of small neighborhoods, away from their suppliers and customers, and to concentrate in urban areas. It relieved the slowness of within-city mail, allowing companies to do more economical business in cities. It allowed the management of manufacturing plants to move away from the factories, putting corporate headquarters in urban areas. And finally, it aided communication within and between large office buildings, making them economically feasible, and thus contributing indirectly to the density and congestion of urban areas.

This permissiveness of early telecommunications technology must not be lost on us today. As more information is available in electronic form, it can be shipped around the world almost instantaneously. The space and time compression of modern telecommunications may have the same mixed effects on social relations and on geographic dispersion as did the telephone. More generally, the flexibility of the new white-collar technologies suggests that uses and effects can be influenced more by other concurrent social changes and by social decisions than by an inexorable technological determinism.

Causes for Poor Social Prediction About Technology

A number of circumstances combine to make prediction of the social effects of technology difficult (Hunt & Hunt, 1985a). First, we have been talking here about the impact of new computer and telecommunications technologies quite generally. But it is concrete realizations of that technology that are implemented, placed into use, and have effects. Word processing can be organized into central clerical pools, as IBM did when it first introduced its equipment in the 1960s, it can be organized using a distributed support staff, in which a clerical person is integrated into a non-clerical workgroup, or it can be organized without support staff, with managers and professionals composing and revising on individual workstations (Johnson et al., 1985). These styles of organization are likely to lead to great differences in the nature of the jobs in which word processing is used, the quality of those jobs, and their number (Benoit, Cossette, & Cardillo, 1984) Thus, to predict the future, we need to assess both the probability that some technology will be used in a particular way and the likely impacts that that implementation will have.

Second, a fundamental fact about innovations is that most fail. Despite the enthusiastic predictions of their inventors and backers, most are not used, and, as a result, have no social effects. The history of white-collar technology is replete with failed innovations. Graham Bell's telephony was a success, but his "triple mouthpiece" for three simultaneous speakers was less so; AT&T demonstrated Picture Phone[1] visual telephone service in 1929 (Ives, 1930) and has periodically introduced it as a commercial service. Although a technological marvel, every commercial incarnation has failed. Of course, the problem for prognosticators is to know in advance which of the prospects will succeed and which will fail.

Third, it is hard to separate successful innovations with effects from those without effects. The problem here is to separate the ballpoint pens from the telephones, innovations with shallow effects from those with profound ones.

Fourth, it is easier to identify destructive changes to the status quo than to identify generative effects. Thus, one can easily identify skills that may become obsolete, groups of workers who may lose their jobs, or industries that may lose business with the introduction of some new technology. More difficult is the task of identifying new skills that will be needed, new classes of workers who will be in demand, and new industries that will be established.

Finally, we have difficulty distinguishing the effects of transition per se from the effects of technology. Change by itself can be disruptive and even more so if the process of change is mishandled. The road to paradise may be rocky. In the case of office automation, for example, an organization can deal with the labor surplus potentially associated with some new technology by firing workers, by reducing its labor force through attrition, or by transferring workers. It can deal with skill obsolescence by firing or retraining. As several chapters in this volume point out, these are social decisions about the methods of transition that can be made independently of the technology itself.

ISSUES IN THE TRANSFORMATION OF WORK

So far we have only hinted at the major issues that the current transformation of white-collar work raises: What effects will the information revolution have on the amount of white-collar employment and on the distribution of employment among occupations, industries, and social groups? What effects will it have on the quality of employment? What effects will it have on produc-

[1] Trade Mark of AT&T.

tivity, broadly defined? And what are the conditions and circumstances that moderate these effects? Can the technology be introduced in ways that meet business goals—enhancing the quality and productivity of white-collar work—and still foster high quality employment? Will some groups, especially clerical women, be differentially affected by the changes that are now taking place? This book addresses each of these issues, concentrating on meeting business goals, on the quality of employment, and on the differential impact on women, and considering the level and structure of employment more briefly.

To answer these questions, we first need to address the reasons that business, government, and other employers are investing in white-collar technologies, because these motives shape the ways that the technologies are used. Employers have three major motives: labor savings, task and product enhancement, and job enhancement. The first reason, labor savings, is the use of technology to substitute machine labor for human labor or to substitute cheaper for more expensive human labor. The second reason for adopting white-collar technologies is product and task enhancement, the production of new, better, or cheaper products and services. This goal often is achieved by extending workers' capabilities, allowing them to perform tasks that were impossible, too difficult, or too expensive without a technological assist. Third, white-collar technology also can be used for job enhancement, to make jobs themselves better, without regard to an impact on productivity.

What has been the influence of white-collar technology as it has been used to achieve these goals? Has it been used effectively to achieve them? Does it have side effects? The answers to these questions are not straightforward. These goals interact and are not always compatible. Technological effects in one organization may have ramifications in other organizations that have not adopted the technology. Effects at the level of the firm may be different when we consider the level of individual workers or the level of the economy as a whole. The following sections outline some issues that must be addressed when answering these questions and provides a framework for the answers attempted in the rest of the book.

Labor Savings and Employment Levels

At the Level of the Firm. A major reason for employers to invest in white-collar technology is to reduce labor costs, either by reducing numbers of employees, by reducing training and skill requirements for jobs, or by transferring jobs to areas of cheaper labor supply. As we have just seen, by some estimates computing time is now cheaper than the labor costs of a white-collar worker. As a result, computers are being used to substitute for white-collar workers. It cut labor costs for turn of the century employers to substitute typewriters and carbon paper for copyists, and for later employers to add ditto- and mimeograph- and then xerographic copier-machines to the arsenal. So too it may make economic sense to substitute word processors and personal computers for secretaries and typists, computers and databases for file clerks, optical character recognizers for human transcribers, decision programs for insurance underwriters, and speech synthesizers for announcers and telephone operators. Certainly, as Iacona and Kling illustrate in their chapter, office automation equipment is sold on the basis of labor-saving promises, regardless of its actual performance.

The question remains whether this use of technology to replace workers has actually reduced numbers of workers in the firms that have introduced it. The answer is yes; firms can indeed reduce the number of employees needed to produce a constant volume of work by introducing computer and telecommunication technology, although the gains are often less impressive than venders claim (see Iacono & Kling, this volume). For example, Bikson (this volume) reports organizations that introduce office automation equipment typically report no changes in staffing levels. For those organizations in which staffing levels are changed, the clerical staff is reduced, but the professional and managerial staffs are increased. In a survey of 112 private firms in Montreal in late 1982 and early 1983, Benoit and her colleagues (Benoit et al., 1984) found that one worker using a word processor could replace 1.8 workers using typewriters. Despite this gain in productivity, the firms in the study reported a 19% increase in secretarial employees since the implementation of word processing. (Also see Gould, 1981, for evidence about the productivity of word processing from the writer's point of view.)

But there is another way in which white-collar technology can save labor costs: by allowing employers to hire cheaper labor in place of more expensive labor. If less experienced, educated and skilled employees using technology could produce the same amount and quality of output as more skilled employees without technology, the employer no longer must pay for the "surplus" experience, education, and skill. *Taylorism,* named for the scientific management strategy advocated and elaborated by Fredrick Taylor (1911/1967), is the traditional method by which managers substitute cheap labor for expensive labor. With Taylorism, an integrated job such as a turn-of-the-century clerk, who might have been a skilled worker involved in all aspects of office work—writing correspondence, arranging meetings, screening visitors, making decisions, supervising staff, taking dictation, and computing accounts—is broken into relatively autonomous components, with different laborers performing the different components. This leads to an overall reduction in labor costs, because typists and receptionists need less training and skill than office managers or accountants and are thus paid less. Clearly, cutting labor costs by fragmenting jobs has implications for the quality of the jobs; this is an issue we examine in the section on job quality.

In addition to its effects on job definition, current technology allows reduced labor costs by moving unchanged jobs to areas of cheaper labor supply. Employers can move electronic data across the country or the world in seconds. One way in which this capability is being used in the airlines, banking, and insurance industries is to move jobs from expensive, northern metropolitan areas to suburban locations, with a large supply of white, well-educated women workers who demand neither high wages nor career opportunities, or to foreign countries, with workers who receive wages below that in the United States (e.g., Noyelle, 1984; Offshore Offices, 1985).

At the National Level. So far we have been considering what happens to employment in individual firms when white-collar technologies are introduced. We must also consider the impact of the white-collar technologies on employment in the economy at large. Many public policy analysts have written about the introduction of technology, both white- and blue-collar, and its effects on levels and structure of employment (e.g., Cockroft, 1980; Hunt & Hunt, 1985b; Leontieff & Duchin, 1984; Rumberger & Levin, 1984). A theme running though much of this discussion is that the effects of technological change on employment cannot be determined in the abstract. The answer depends on the interplay between changes in productivity, changes in demand, and changes in other historical and social factors.

In particular, the level of employment for a particular occupation is based on the health of the general economy (general economy), the employment levels in particular sectors of the economy that heavily employ that occupation (sector growth), and the differential hiring and retention of occupational members (staffing ratios) within sectors of the economy (Hunt & Hunt, 1985b). For example, the overall level of clerical employment in the economy will be based on the timing and severity of the next recession, the continued growth or decline of industries that employ large numbers of clerical workers (e.g., banking, insurance, and government), and the actual hiring of clerical workers within those industries. New technology has the most direct influence on clerical employment by influencing staffing ratios, the differential hiring of clerical as opposed to other occupations.

It is an economic truism that in the production of a particular good or service, if technology increases the productivity of labor and demand for output does not increase, then employment must fall—immediately or eventually—in the innovating firm or another, in the number of workers or number of hours worked, in workers fired or in workers not hired. The drastic decline in employment in agriculture, from over 50% of the labor force in 1880 to under 3% in 1980, can be traced to changes in farming technology—machines, chemicals, seeds, and farming practices—in the face of only moderate increases in demand. In addition to direct displacement of farm workers, the productivity gains in agriculture contributed to the shortening of the average farm work week.

But technological increases in productivity need not necessarily lead to declines in employment, either absolutely or relatively, primarily because

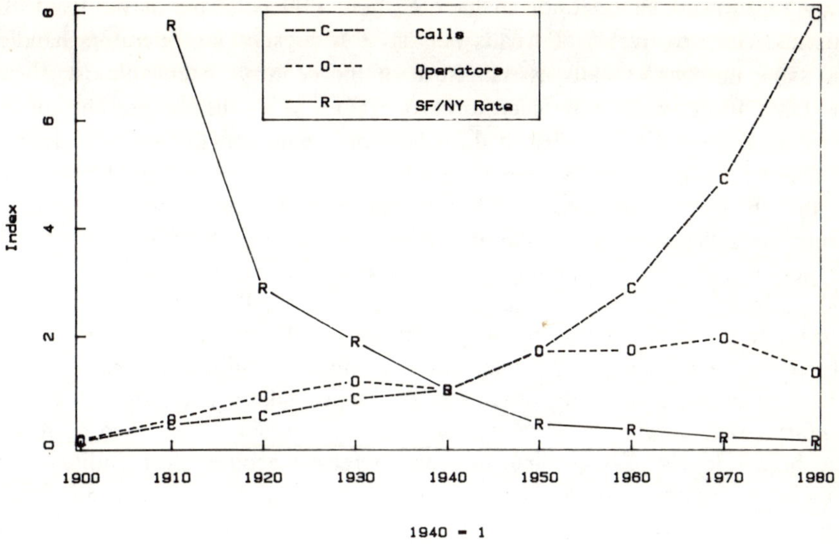

FIGURE 1.1. Telephone use, price, and employment.

demand for output often increases as well. A standard argument in discussions of the employment impact of automation is that increased productivity in one sector often leads indirectly to increased employment in that or other sectors (e.g., Sadler, 1981). Thus, the information technologies that threaten clerical jobs also have opened up new employment opportunities for clerical workers by making the collecting and processing of information economical. Indeed, the very occupation of keypunch operator, which may be automated away with such advances as optical character recognition or direct customer input, and the job of telephone operator, whose numbers are decreasing with the increasing use of electronic switching machines, customer information databases, and electronic speech synthesis, were themselves creations of automation.

Again the telephony field provides interesting examples. The Bell System and other telephone companies since their beginnings have been heavy users of technology to increase the quality of phone service and the productivity of labor. Figure 1.1 plots the number of telephone operators, the number of telephone calls, and the price of a New York to San Francisco telephone call from 1910 to 1980 in constant dollars, using 1940 as a base year (U.S. Bureau of the Census, 1975, Series D233–682, R1–12, R13–16; U.S. Bureau of the Census, 1984, tables 692 and 948).[2] In 1900 technology and work pro-

[2] The price figures were projected backwards, because long distance service between these cities did not start until 1915.

cedures allowed an operator to handle about 400 calls per day. By 1982, this had risen to over 2,800 calls per day. If present-day operators handled the same number of calls as their turn of the century sisters, almost 2% of the labor force would now need to be operators to handle the present volume of calls compared to the .3% of the labor force who actually are operators.

Clearly, technology decreased employment from what it would have been in the absence of technology. But technology also decreased the price of telephone service, increased the quality of service (e.g., the New York to San Francisco call took 14 minutes to effect in 1920 and less than 30 seconds today; voice and connection quality are better as well), increased the ease of service (e.g., direct dialing for long distance calls), and increased the number of services available (e.g., private lines, conference calling, centrex service). These features, along with other social changes taking place during the last century, for example the urbanization and then the suburbanization of America, higher levels of disposable income, increased geographic mobility of the American population, and declines in the extended family have all contributed to an increased demand for telephony services. This increased demand led to an absolute increase in the number of employees of telephone companies from the beginning to the present day and prevented the absolute decline in the numbers of telephone operators from occurring until the 1970s.

While the interplay of technology, productivity, and demand also have informed earlier discussion about the employment impacts of automation, some argue that the current economic environment is so different from earlier ones that earlier analyses are irrelevant (see Chamot, this volume). For each of the last three economic recoveries, structural unemployment (the level of residual unemployment during the period of maximum employment) has increased. With 7 or 8% structural unemployment, the economy may not be able to absorb some of the workers displaced by or not hired because of automation: some segments of the population—the old, minorities, unskilled, or uneducated—may not participate in the benefits of productivity. Compounding this problem, the current wave of automation may demand such levels of skill, general training, and credentials that a substantial minority of workers may be locked out of the labor force (Noyelle, 1984).

Employment Quality

Earlier discussions of the social impacts of automation have focused primarily on the numbers and types of jobs automation would replace or create (U.S. National Commission on Technology, Automation, and Technological Change, 1966). Today the debate rages around the effects of automation on the quality of employment as well. Commentators ask whether white-collar technologies lead to the fragmentation of work or to its integration, to the deskilling of work or to high skill demands, to electronic monitoring of employees or to greater job autonomy, and to the exploitation of workers with limited job opportunities or to freedom for workers to work at times and places convenient for them. Analysts wish to establish whether white-collar technology is used primarily to ennoble or degrade jobs and to establish the conditions under which it does one or the other.

The Nature of Employment Quality. To answer these questions, of course, we need some criteria for assessing the quality of employment. This, in turn, depends on the perspective from which we are defining employment quality. At a minimum, employers, who may want to optimize productivity, product quality, and worker dedication, undoubtedly have different standards than employees, who may want to optimize intrinsically interesting work, compensation, job security, or pleasant working conditions. Among themselves, both employers and employees also may have different standards, depending on their social circumstances, group memberships, and individual preferences.

It is true that to some degree employment quality is subjective and idiosyncratic and reflects the fit between particular worker's needs and job characteristics. Still, the literatures on job satisfaction, job performance, and job design suggest at least three categories of attributes that influence employment quality for most workers (Hackman & Oldham, 1980; Locke, 1976). These are the job content, compensation, conditions, especially the social conditions, under which the job is done.

Job content encompasses those attributes that are intrinsic parts of the work. There are a number of job attributes—summarized under the rubric of challenge, especially mental challenge (Locke, 1976)—that increase employment quality. Learning, creativity, autonomy, responsibility, variety, and coping with difficulties exercise workers' conceptual facilities, while the lack of these qualities produces boredom. When technology is introduced in ways that enhance these qualities of jobs, the jobs get better for most workers, as witnessed by an increase in workers' satisfaction and their commitment to their job (Hackman, Pearce, & Wolfe, 1978).

Compensation includes both the absolute level of wages and salary, fringe benefits, security, and promotability that a job offers, and the fairness of these factors compared to norms of an occupation and industry and the inputs that a worker provides.

Working conditions are both physical and social. Dangerous and uncomfortable conditions decrease employment quality. The precise conditions that make a job dangerous or uncomfortable can vary, and include limb-threatening or vision-threatening equipment (National Research Council, 1983) as well as extremes of temperature, noise, lighting.

Workers get satisfaction from being with other people on the job. On-the-job friendships are especially important sources of social satisfaction and support for white-collar workers because communication is often an intrinsic part of their jobs (Panko, 1984) and because they spend so much of their time among work colleagues. Social satisfaction and support from co-workers are among the reasons that people who are employed have better mental health that those who are not (e.g., Thoite, 1983).

Job Fragmentation and the Deskilling Debate. How is white-collar technology associated with these sources of job quality? The major portion of this book deals with just these issues. The following discussion is an attempt to put this work in perspective by outlining some important issues about technology and job quality.

The possible impact of white-collar technologies on the content of jobs has been most elaborated. Braverman's (1974) seminal work has shaped the terms of this debate. Adopting a Marxist-critical approach, he argued, as just discussed, that the introduction of technology into the office is a mechanism to rationalize and fragment office work. As in the factory, office work is broken into many subtasks, each performed by a "detail" worker, who loses the integrative contact with the total product and who loses variety in the job. The consequences from the employer's perspective are to reduce the skill requirements of office work, that is the average employment experience and education needed to adequately perform the job, and, therefore, to reduce labor costs. The consequences of job fragmentation from the employees' perspective are to reduce the quality of the job, by reducing the variety of tasks they perform, mental challenge, and job autonomy and responsibility.

Within this framework, research has had two major goals. First, it has attempted to document changes in employment quality in newly automated white-collar work. And second, it has attempted to establish the casual role that technology has played in these changes. The conclusions from this research are contradictory. I argue that some of the contradictions result from the study of technology at different epochs and stages of maturity. In addition, the contradictions reflect comparisons across levels of analysis and a basic ambiguity about definitions of skill and deskilling.

Most investigations into the skill requirements of white-collar technologies have been case studies of particular firms or industries. Researchers have established in a number of cases that the introduction of white-collar technology coincided with the fragmentation of jobs. For example, Murphree (1984) documents the fragmentation of the job of legal secretary coincident with the introduction of word processing equipment, although the role of technology in effecting these changes is unclear. The variety of tasks that made the job of legal secretary both interesting and highly paid were stripped from it. Legal aids composed first drafts of simple contracts, specialized legal librarians searched for legal citations, and messenger services hand delivered urgent documents, although these tasks had previously been done by legal secretaries. Even the task of producing a manuscript was decomposed into an initial typing stage and a later proofreading stage, with different personnel performing each function. In a different example, when AT&T introduced computerized record systems for its customer service representatives during the mid and late 1970s, it split the universal service representative's job in two. In the old job, skilled clerical workers had responsibilities for solving customers' problems, selling equipment and service, and collecting debts. After the split, the service representative's job becomes more specialized to deal either with service orders and sales or with debt collection. One motivation for this specialization was to reduce training costs; a consequence was to reduce task variety and thus make the jobs less interesting. (cf. Feldberg & Glenn, this volume, 1977; Maddox, 1977).

What role did the technology itself play in these and similar changes in job definition? Scholars are now debating whether new technology encourages job routinization and fragmentation, is a beneficiary of it, mitigates it, or is independent of it. Certainly, scientific management of office work was occurring well before the current introduction of microelectronics (cf. Davies 1983; Leffingwell, 1925; Murolo, this volume). Some have argued that job fragmentation and routinization results from the inherent conflicts of interest between capitalists and workers (Braverman, 1974; Driscoll, 1980) and others that it is a bureaucratic response to the increase in office and paper work resulting from, for example, increased sizes of firms, national markets, or government record keeping requirements (Murphree, 1984; Strom, 1985). Finally, others argue that fragmentation and routinization are necessary preconditions for the application of the simplistic computer technologies that are available at the current stage in the automation of office work, but are not necessarily consequences of the automation (Baran & Teegarden, 1984). Work that has already been Taylorized is a prime candidate for automation.

In contrast to this range of views that see technology as a tool by which work is rationalized, other analysts have presented theory and data suggesting that new office technologies can reintegrate jobs that had been previously fragmented for other reasons (Matteis, 1979; Strassman, 1980). At the clerical level, for example, by using centralized databases, customer service representatives can handle all of the transactions associated with a client's account, taking orders, entering data, making adjustments, and answering inquiries. (Baran & Teegarden, 1984; Feldberg & Glenn, this volume). At a more professional level, using new workstations, managers can have greater control over more stages of their work by using computer-based tools for searching corporate databases, calculating the consequences of alternative investment strategies, creating illustrations, outlining, writing, spelling checking, formatting, and sending the resulting report to a mailing list of recipients (Spinrad, 1982).

The conflicting views on the impact of technology on work routinization and fragmentation are paralleled by conflicting views on the skill content of work. In general, those researchers who believe that automated jobs are more fragmented than nonautomated ones also see the automated jobs as requiring less skill and education (e.g., Driscoll, 1980; Nussbaum & Gregory, 1980). According to this view, software incorporates substantive knowledge of a job and its procedures, an establishment, and an industry, leaving less for the worker to know. For example, supermarket clerks using computerized cash registers with bar code readers are not mentally challenged to recognize brands and promotions, to memorize product-price lists, or to perform mental arithmetic; that is, they need to know less about the food marketing business than their non-automated counterpart of 20 years ago. On the other hand, starting from the assumption that the most routinized jobs in an industry are the first to be automated, Adler (1984) argues that automation increases skill requirements for those jobs remaining, especially generalized cognitive skills associated with computer use, increases worker responsibility, and increases co-worker cooperation. For example, if most routine banking operations are performed by customers themselves using automatic teller machines, the banking tasks remaining to human tellers are non-routine, e.g., handling problem inquiries or selling specialized services, that may require considerable knowledge of banking procedures, problem solving skills, and skill at dealing with people.

Stages of Technology. In part, these disagreements in the research literature about the effects of technology on job routinization, the skill requirements of jobs, and the skills of workers are the result of comparisons of experiences at different stages in the evolution of the technology itself and of different social and economic circumstances under which the technology is introduced (Baran & Teegarden, 1984). Given the rapid changes in the technology itself and in the experiences using it, conclusions from one wave of technology use may not generalize to later waves.

A examination of the social organization of word processing is a case in point. Large word processing pools are much more likely to be the modal organization of word processing personnel if the word processing equipment is old (pre 1978) rather than new (post 1981) and if the firm introducing word processing is large rather than small (Benoit et al., 1984). Women who work in large pools have more homogeneous and boring work than do other word processing secretaries. They spend over 90% of their time doing typing tasks, while those who work in mini-pools serving several clients or who work as private secretaries for a single principal have much greater variety in the tasks they perform (Benoit et al, 1984; Panko, 1984). It is not clear whether these differences in the social organization of word processing reflect the differing technical capabilities of old rather than new equipment (e.g., minicomputer-based vs. microcomputer-based word processors) or changes in management philosophy resulting from experiences with previous implementations (e.g., a "learning curve") or other social influences. But the point remains that researchers trying to understand the role of technology in job fragmentation would have come to different conclusions if they had studied large or small firms using old or new equipment.

Levels of Analysis. Second, the disagreement in the research literature results because researchers have ignored important level-of-analysis distinctions about the units of work that are being skilled or deskilled (cf. Feldberg & Glenn, this volume). The skills and skill requirements of tasks, individuals, jobs, occupations, firms, industries, and the economy as a whole can change with the introduction of technology, but a change at one level has no necessary implication for changes at other levels. A particular task may require less skill to perform when it is done on a computer, but depending on the variety of tasks comprising the jobs and the job holder's entering knowledge, the job itself may become more satisfying and challenging.[3] This paradox may be compounded as tasks migrate between jobs and occupations and as new generations of workers fill the transformed jobs. Perhaps some examples will make this point clearer.

First, let us consider the introduction of a word processing system among a group of secretaries. The word processing system allows text editing of manuscripts, formatting (e.g., centering, underlining, justification, pagination, and setting fonts and point sizes), sorting, and syntax and spelling checking. As a direct result of the word processing system, the skill and educational requirements needed to perform some specified tasks have decreased. In order to sort the entries in a bibliography, a secretary needs less facility with the alphabet, and to proofread the manuscript, she needs less knowledge of spelling and grammar.

In addition, some of the procedures a secretary performs to achieve a certain result have changed (e.g., centering or underlining text), although we cannot determine in advance whether the change in procedure represents an increase or decrease in skill requirements. Finally, some tasks the secretary performs in creating an electronic manuscript have no counterpart in a typewritten manuscript; thus, for example, setting the size and style of font is more similar to traditional typesetting operations than to typewriting ones.

---

[3] This discussion proceeds as if the skill and education needed to perform a job were solely a function of the tasks comprising the job. Clearly, however, the skill requirements of a job have political and social foundations, as well as technical ones. For example, the increased educational requirements for secretaries found in employment advertisements over the last 30 years may reflect the supply of college-educated women and the association of education with other class and racial characteristics that employers find desirable more than they reflect changes in the secretarial occupation.

It is not clear in the abstract what the net effect of these changes in tasks will be on the net skill required to produce a manuscript. Let us assume for the moment that overall, the training, skill, and education needed to create a manuscript using a word processing system is less than that required using a typewriter, although, the skill requirements may in actuality be increased. There is, however, no necessary connection between this changed skill requirement and the skill exercised by particular secretaries, the skill requirements of the job of secretary within a particular establishment, or the skill or educational requirements of the secretarial occupation as defined in private and public occupational classifications.

The skill requirements of jobs depend on the way that tasks are organized to form jobs. Some jobs may be organized so that they are composed only of tasks that have been deskilled by technology. This organization seems to have been the case, for example, in the word processing section of the large law firm that Murphree (1984) studied, where the typing and proofreading functions were separated. On the other hand, the secretarial job can be organized, as it is in many universities, so that secretaries perform many services for the principals for whom they work. In this case, where the technology is introduced into a job involving a mix of tasks, the increased speed at performing some tasks may lead to increased time and energy available for other tasks, including, in the case of university secretaries, gate keeping, student advising, monitoring grants and contracts, and administering student records. This state of affairs is likely to lead to skill enhancement for the job as a whole, if, as is often the case, the most routine tasks within a job are automated first.

Again, the amount of skilling or deskilling of tasks and jobs has no necessary implication for the skills that a particular job incumbent can use. Clearly, because procedures of work have changed, some of a job incumbent's knowledge will have become obsolete whereas some procedures will be new and remain to be learned. But knowledge change is not equivalent to deskilling. In the case of an individual, a deskilled job is one that fails to exercise general mental or physical capabilities and is not challenging. Whether a job is challenging is a function of both the skill level required by the job and the skill level brought to it by the incumbent. Thus, if a job were deskilled (i.e., the sum of its tasks required a lower level of skill, training, or general education) the challenge it offered job holders would depend on their entering skills, training, and education. When the same people hold the jobs before and after the introduction of technology that lowers the skill requirements of jobs, they are unable to use their general knowledge and experience, and their jobs become less challenging and satisfying (e.g., Rogers & Friedman, 1980). On the other hand, the introduction of technology that lowers the skill requirements of a job often affects new job entrants, not old job incumbents. Job incumbents often retire, or are promoted, transferred, or laid off (e.g., Rogers & Friedman, 1980). The new job entrants—often with less employment experience and education—may well be challenged by the level of work. Indeed, the deskilling of some white-collar jobs may be the vehicle by which less advantaged social groups gain white-collar work.

In addition to information technology changing the skill level required to perform certain tasks, its introduction is often the occasion for the reorganization of existing tasks across job categories. Thus, in the law firm that Murphree (1984) studied, the shifting of library duties such as tracking down legal citations from legal secretaries to specially trained legal librarians working with sophisticated computerized data systems was one element in the the secretaries' deskilling. According to Murphree, the legitimacy of billing for para-professional services was a more important cause for this change than were technological factors. In universities, professors do much of their own typing with word processing equipment and personal computers, transferring some of the most onerous of data entry tasks from secretaries to professionals while at the same time giving the professionals more control over the form and content of their writing. Especially in the case of the transfer of tasks across jobs, we need to recognize that shifts in the skill requirements of one job may be associated with complementary shifts in the skill requirements of another.

So far we have discussed the skill requirements of jobs, i.e., the bundle of tasks held by particular people in particular organizations. We have seen that the deskilling of some tasks does not necessarily imply a deskilled job, depending on the variety of tasks comprising the job, the creation of new tasks, and the shifts of tasks between jobs.

Occupations are the aggregation of similar jobs across the economy. Determining whether an occupation has been deskilled entails similar problems to determining whether a job has been deskilled. In addition, one must contend with shifts in the definition of the occupations. In particular, the U.S. Bureau of the Census revises its occupational classifications periodically and adopted a new classification of occupations starting with the 1980 census. The U.S. Bureau of Labor Statistics adopted a new classification of occupations to be used in some of its time series data starting in 1983. These changes have made all comparisons across time, for both the amount and quality of work, extremely difficult.

Transition. Whatever the net effect technology has on skill requirements, there will be some jobs that require less skill and experience after automation than before while other jobs will require more. The changing skill requirements associated with automation, regardless of the direction of change, are bound to produce discrepancies between the skills that job incumbents and entrants have and the skills the jobs require. These discrepancies are a problem to contend with in their own right.

Other Quality of Work Issues

Abstraction of Work. Zuboff (1982) argues that in addition to the effects of technology on the organization of and conditions of work, new office technologies change the texture of the work itself through computer-mediation. She argues that both clerical and managerial workers are distanced or alienated from the information they deal with because they manipulate only electronic shadows. To counter-act this tendency, statistics professors may advise their students to perform complex analyses by hand rather than by computer, because the first-hand contact with their data gives them a richer appreciation of their findings. Similarly, some small publishers compute their royalty statements by hand to gain a more immediate appreciation of which books are selling. Analogously, Kiesler (1984) finds that computer-mediated communication distances the people writing to each other and is emotionally cooler and less personal than face-to-face communication (cf. Short, Williams, & Christie, 1976).

If this alienation from information is a large effect, some of it may be caused by the type and amount of information presented to workers and the style with which it is presented. In a batch style of information processing, characteristic of large computers of the 1960s and 1970s, a manager or other worker may wait hours or days between requesting information and getting it. This batch style may be much more distancing than highly interactive computer systems of the 1980s. In these interactive systems, information exchanges happen in real time. A secretary can see the effects of a change in a margin, a draftsman can erase and redraw a line, or a manager can recalculate interest payments instantly.

Ergonomics. Finally, in the area of working conditions, the issue of the effects of video display terminal use has received much attention. One of the characteristics of the new white-collar technologies is that many workers, both clerical and professional, spend much of their work time sitting in front of a video display terminal. Some of the discomfort associated with prolonged terminal use (Nussbaum & Gregory, 1980) can be traced to traditional human factor decisions about lighting, screen legibility, seat design, and keyboard design; recent research and design standards are addressing these problems (see Dainoff, 1982; Helander & Rupp, 1984; National Research Council, 1983; M. Smith, 1984 for overviews; see also Bjørn-Andersen & Kjeargaard, this volume).

Need to attend to the ergonomics of the office may be nothing new. Melville's scrivener of the 1850s found working conditions difficult; for example, he showed

> continual discontent with the height of the table where he worked. . . . He put chips under it, blocks of various sorts, bits of pasteboard, and at last went so

> far as to attempt an exquisite adjustment, by final pieces of folded blotting-paper. But no invention would answer. If, for the sake of easing his back, he brought the table-lid at a sharp angle well up towards his chin. . . , then he declared that it stopped the circulation in his arms. If now he lowered the table to his waistbands, and stooped over it in writing, then there was a sore aching in his back (Melville, 1856/1959, p. 6).

But the 19th century copyists did not have to contend with the potential health risks associated with video display terminals per se. Although a National Research Council report (1983) concluded that these terminals pose no long-term risks of vision damage or long-term radiation risks, the opposing views (e.g., Nussbaum & Gregory, 1980; Stark's dissent in the NRC report) make further research necessary.

CONCLUSION

In the midst of the current revolution in white-collar work, it is hard to see the current trends and to set standards. Certainly we need continued research to understand how to design and deploy computer and telecommunications technology to be both productive, in its widest sense, and humane. This technology is flexible. Given this flexibility, we can have many arrangements of work with only minimal technological constraints. The technology also is rapidly changing. Given the rapid change, current solutions may not hold in the future. These considerations suggest that in addition to the search for the "correct" solution to the design and deployment of the technology, we must be sensitive to the processes by which the technology is designed and deployed. Perhaps a process of design, implementation, and use in which the affected interest groups—employers, workers, and consumers—can influence the final decisions about the way technology is used is more important than the quality of any single decision or standard.

2 Electronic Work and the White-Collar Employee

Dennis Chamot
Department for Professional Employees, AFL-CIO

ABSTRACT

This chapter addresses concerns that organized labor has about the introduction of white-collar technology. In terms of employment prospects, advanced office technology currently threatens employment for middle managers and clerical workers and eventually will do so for professionals as well. Given the increasingly high unemployment in the past decade, labor growth in new industries and occupations may not compensate for the labor reductions that are associated with new technology. In terms of job quality, use of white-collar technology can lead to deskilling and rationalization of work, increased stress associated with poor ergonomics, computer pacing of work, job speed-ups, excessive monitoring of work by computers, and the exploitation of workers who work from home using computers and telecommunications equipment. These concerns about job security and working conditions result from choices made by equipment and software designers and by managers. Many of the problems, however, can be avoided if employees are actively involved in early stages of the change process.

The technological transformation of white-collar work is occurring within a context of rapid change in all areas of employment. Computers, microchips and new communications technologies are having profound effects on how work is done in factories, warehouses, retail outlets, and virtually every other type of employment facility in the United States.

The last few years have seen the introduction of a growing array of new office technologies based upon small computers and complex communications systems. Their effects on office workers are only beginning to be felt.

To properly analyze these effects, I find it useful to subdivide the white-collar workforce into three categories:

1. Executives and managers—among other things, these people have supervisory authority over subordinates; they set policies, and see that they are carried out. Middle managers, in particular, devote much effort to oversight, data gathering and report production.

2. Professional employees—these workers usually have college degrees, and have jobs that involve frequent use of judgment and the application of specialized knowledge.

3. Clerical and other support staff—this group includes secretaries, typists, copy machine operators, and the like. For the purposes of this chapter, *clerical employees* refers to essentially all office staff who are neither managers nor professionals.

I refer to this division later, because modern office technologies are affecting each category in somewhat different ways.

EMPLOYMENT EFFECTS

Let us begin by discussing employment levels. This is one of the most difficult issues to deal with. Nevertheless, it is certainly one of the most important from the employee point of view. To put the question in proper perspective, it is useful to look at a little history.

In the early days of this nation, the bulk of the workforce were engaged in food production (see Table 2.1). Today, that portion is down to about 3%.

TABLE 2.1
Employment in Food Industries

	1900		*1982*	
Total Population (millions)		76.0		231.5
	No.	% of workforce	No.	% of workforce
Farmworkers	10.9	14.3	2.7	1.2
Nonfarm Food Related Employees:				
Food Processing	0.1		1.6	
Food Wholesaling	0.3		0.8	
Food Retailing	0.4		2.5	
Eating and Drinking Estab.	0.1		4.8	
Total Nonfarm Food Related	0.9	1.2	9.7	4.2
Support Activities (farm equipment, fertilizers, hauling and delivery)	0.2	0.3	0.9	0.4
Total Food Related Workforce	12.0	15.8	13.3	5.8

Adapted from Robert S. Levi, in *Manufacturing Productivity Frontiers,* July 1984, p. 22.

In fact, if one were to look at all aspects of food production and sale (farming, fertilizer and tractor production, food processing and retailing, and so on), about 12 million people were engaged in this industry in 1900, and only about 13 million are today, even though our population tripled and the industry underwent massive change (e.g., the explosive growth of fast food establishments, see Table 2.2). Clearly, farms today are enormously more productive than they were in the past, thanks largely to technological developments.

A similar trend has been occurring in manufacturing. The number of production jobs has been stagnant for years. What job growth has occurred has been in white-collar areas, so that on average, one third of the jobs in manufacturing firms are now white-collar jobs. The widespread redesign of production facilities around robots and other computer-driven equipment over the next decade or two will undoubtedly result in further erosion of the blue-collar job base, and this result can occur even with some expansion in manufacturing output.

Now we are beginning to see similar developments in the service sector. Telecommunications, of course, has been dependent on sophisticated technologies for a long time, but that is an industry that existed in a protected environment and grew rapidly. New services may offer new job opportunities, but in the face of stiff competition, there also is considerable pressure to reduce labor costs through increasing automation.

The explosive growth of computer usage, coupled with rapid data transmission, is leading to a melding of the banking, insurance, and financial services businesses. Predictions are that total employment will drop, and it is easy to see why. Firms in the insurance industry already report a decrease in overall employment, attributed primarily to the introduction of electronic office technologies over the past decade (Appelbaum, 1985). In banking, computers are being used both to replace tellers (through automatic teller machines) as well as to increase productivity of "back room" clerical and accounting operations. One result will be reduced employment opportunities, even with economic growth. Such is the prediction of banking officials as well as outside observers.

TABLE 2.2

U.S. Manufacturing Employment, in Thousands, 1960–1985

Year	*All Employees*	*Production Workers*
1960	16,883	12,659
1970	19,603	14,246
1980	20,146	14,047
1985 (March)	19,713	13,465

Data for June of each year.
Source: U.S. Department of Labor, Bureau of Labor Statistics.

In the past, loss of job growth in industries undergoing automation was largely balanced by opportunities in other industries that were not automating as rapidly, or where growth was occurring. For the past couple of decades, much of our employment growth has occurred in the service sector, including government. However, it is precisely this area—predominantly white-collar and office work—where the latest round of automation is just getting underway. Where will displaced people go? We have run out of sectors (see Fig. 2.1).

Some would argue that increased economic growth will create enough new jobs to provide opportunities for workers who are displaced by new technologies. The fact that this may well not happen is dependent on the observation that employment is dependent on the relative rates of growth of both productivity and output. In a straightforward way,

$$\text{productivity} = \text{output/hours worked.}$$

If we consider total productivity, then we can replace hours worked by total employment. Rearranging terms gives

$$\text{employment} = \text{output/productivity.}$$

It is easy to see, then, that if productivity growth is rapid and output increases but at a slower rate, not at all unreasonable assumptions, then employment can drop.

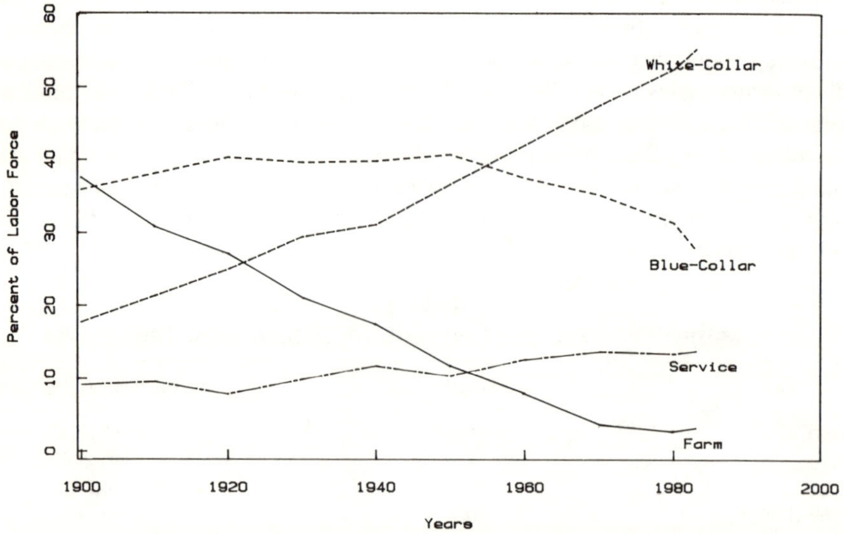

FIGURE 2.1. Changing composition of the U.S. workforce, 1900–1983.

We have witnessed a "ratcheting upwards" of unemployment in this country—the rate at the peak of an economic expansion (the lowest unemployment rate to be expected during any given business cycle; see Table 2.3) seems to be increasing from cycle to cycle.

Various explanations have been offered to explain this behavior. One recent study (Podgursky, 1984) examined the effects of an influx of young workers (the "baby boom" generation) and a rising labor force participation rate among women. Both were important, but at different times. According to Podgursky, "Both the peak-to-peak and trough-to-trough measures suggest that the preponderant reason for the rising unemployment rate was *involuntary job loss*" (italics added, p. 22). The reasons for this increase are still speculative. In very recent years, the very high value of the dollar undoubtedly plays a major role. I would submit that we also are witnessing the early effects of the massive shift to office automation and computer-aided design/computer-aided manufacturing (CAD/CAM) that began in earnest during the past decade or so. Without enormous economic expansion or major changes in public policies or a much slower introduction of the new systems that most people think will happen, we may well develop a permanent labor surplus (AFL-CIO, 1983).

Now it is time to go back to the workforce subdivision mentioned earlier, because the employment effects are not going to be identical for each of the three groups. I believe that middle management will be strongly affected. The jobs of many middle managers involve employee oversight, data gathering, report generation, and general coordination—all functions that, to a greater or lesser extent, can be performed with computerized networks and fewer people. For example, loading dock workers now use keyboards or light pens and bar codes to record unloading of parts and raw materials and shipments of products; reports related to shipments and material in inventory can then be obtained quickly and fairly automatically. Similarly, supermarkets need not do involved inventory counts any more; the scanners at the check-out

TABLE 2.3

Average Unemployment Rates at Business Cycle Peaks and Troughs, 1969–1982

Peak Quarter	*Unemployment Rate*	*Trough Quarter*	*Unemployment Rate*
1969: III	3.6	1970: IV	5.8
1973: III	4.8	1975: I	8.3
1979: IV	6.0	1980: III	7.5
1981: III	7.4	1982: IV	10.6

Adapted from "Sources of Secular Increases in the Unemployment Rate, 1962–82," by Michael Podgursky, *Monthly Labor Review*, July 1984, p. 20.

counters do this automatically. In various work environments today, sophisticated software packages take a lot of the work out of scheduling and maintaining attendance records. In these and many other cases, computers have been used to assume many of the more routine functions usually performed by middle managers. Not every business is at a high level of sophistication in this area, but as software improvements occur, applications in this area are bound to increase.

Professional level employees present a more complex situation. Frequently, they see the changes as improvements in their tools, offering greater speed or flexibility in the performance of their duties. This is certainly true, for example, for engineers who use computer-aided design systems, or for economists who have their computers do endless calculations and prepare large numbers of charts and tables. However, the very characteristic that is so useful to individuals—much greater speed and increased capacity—may well have a strong negative effect on future job demand. There also is the problem of downgrading—jobs that formerly required professional training, but now, because of sophisticated yet user friendly software packages, can be performed satisfactorily by lower level employees.

Of course, whole new fields also are being created—programmers and systems analysts, for example. The question is whether the rate of job creation will equal or exceed the rate of job consolidation and elimination. No one knows.

The third group—clerical and other office workers—will experience a large reduction in demand over time. Much office work is quite routine. Many people are needed to move paper around, to type, file, retrieve from files, copy, distribute, retype, cut and paste, amend, and so forth. The so-called electronic office will allow the generator of the information or report to do a lot of these things directly, eliminating the need for many of the support people.

Clerical employees also are being presented with opportunities for upgrading. Many of the downgraded jobs referred to in the discussion of professional employees will, in fact, be filled by, and represent upgrades for, clericals. Even in those cases (and only time will tell if the net result is an overall employment decline across *all* groups) the redesigns often result in dead-end jobs with no further promotional opportunity (Appelbaum, 1985).

White-collar workers—clericals, professionals, sales people, and managers—currently constitute over half of the entire American workforce. Clericals are the largest segment, accounting for one out of six people in the workforce. This group has been one of the fastest growing for years. Managers represent another 10%. Between them, managers and clericals currently constitute a quarter of the workforce. And they are the ones whose jobs are most at risk with the advent of office automation.

CHANGES ON THE JOB

Not everyone will be without work, of course. Most will be employed, and new office technologies are already changing the way they work. There are many positive developments, e.g., word processing; computational and graphics software for business and personal computers; rapid access to electronic data bases; electronic mail; computer searches.

There also are some negative developments. An important one is deskilling. This is already happening in the insurance industry. Computer storage of vast amounts of information, as well as standard forms and risk criteria allow people with lower levels of skill than were required in the past to sell insurance policies, while older agents lose their skills as they become more dependent on the computers.

The development of artificial intelligence (AI) programs and expert systems will undoubtedly have a profound effect on skill requirements for even highly sophisticated kinds of work. Although still in their infancy, AI expert systems can do a reasonably good job, for example, in certain areas of medical diagnosis, and in assisting geologists in locating ore deposits. In the beginning, such approaches may be designed to assist professionals in making better or more accurate decisions, but it is easy to see how in the future artificial intelligence can be used to try to replace many professional level employees with lower skilled (and less expensive) technicians.

Along with deskilling goes job rationalization. It is interesting that developments in office automation are in many ways following the traditional factory pattern, even, ironically, as manufacturing firms seek new, more humane approaches. Typing pools return, to try and squeeze maximum "productivity" out of expensive word processors; other jobs are broken down into simple, repetitive activities, such as endless keyboarding. Interesting secretarial jobs with a lot of variety are replaced with boring, repetitive, dead-end jobs.

Again, *deskilling* and *upgrading* are, in a sense, relative terms. Although some employers are using new office technologies to redesign work organization around rationalized job functions, others are taking the opportunity to broaden clerical jobs and to increase the variety of tasks performed by an individual. In such cases of job expansion, the clerical employee who performs the new work experiences an upgrade. However, if some of these functions used to be performed by a higher level employee, then that employee sees essentially the same job as being downgraded.

Increasing stress also seems to be a characteristic of electronic work, caused primarily by three factors: poor ergonomics, computer pacing, and excessive employee monitoring.

Although there is still some dispute over harmful effects from non-ionizing radiation emitted by visual display terminals, there is fairly wide agreement among professionals in the field that better design of equipment and work surroundings, as well as rest breaks, are needed to reduce stress and relieve muscle aches and other physical ailments. Many unions are actively working in this area, both in this country as well as in Europe. Redesign of work locations and the institution of regular rest breaks are key areas where joint management–union activities would be of great value at specific work sites. Unfortunately, such cooperative approaches in the United States have not been as successful as they should be, and thus many unions are supporting protective legislation in various states as well as at the federal level.

Computer pacing is a problem, and bitter complaints are heard from telephone operators who are instantly assigned call after call from a computer dispatcher, and from data processors who are assigned yet another form on their screen the instant work on the old one is finished. The problem, of course, is that people are not machines. No one can easily work at the same pace, without break, hour after hour, on what to them is often boring, repetitive, meaningless work. This is job design at its worst.

Electronic work is very easy to monitor. No one likes to have somebody constantly looking over their shoulders. You want to be able to make mistakes, as everyone does, and correct them, without fear that management will see them before you do, and draw the wrong conclusions. Most people appreciate the ability to vary work pace when possible, working rapidly when fresh, slowing down a bit when tired, without fear that a supervisor will check at the wrong time and draw the wrong conclusion. There is something very disquieting about the fact that an impersonal machine is running your life so directly.

Swedish employers are not allowed to use computers to monitor the work output of individual employees, yet work seems to get done (for more on Swedish approaches to workplace technological change, see Chamot, 1985). They have even been able to experiment quite a bit with forms of work organization designed to give more autonomy and decision making authority to individual workers. That is very different from what has become a fairly common technique in American offices, where computers are used, for example, to count keystrokes. Such an emphasis on measuring individual "productivity" frequently gives poor results. This should not be surprising; if the emphasis is on hitting a lot of keys, then the operator will not have the time to make corrections, and expensive errors and procedures for correcting them will increase.

HOMEWORK

There is another aspect of electronic office work that has a serious potential for abuse, and that is its portability. This very characteristic is touted as a major advance for professional and managerial employees who do some extra work at home, or who travel frequently. However, the situation is very different for clerical workers who work out of their homes and do not share the same rewards and perquisites of other office workers who occasionally choose to work at home for personal convenience.

Homework is not a new phenomenon. Historically, many industries used the services of homeworkers—sewing trades, costume jewelry makers, button makers, and others. Abuses were rampant, and included long hours for minimal pay, use of child labor, and unsafe or unsanitary working conditions. Numerous attempts were made in several states to regulate homework and thus minimize the abuses, but these efforts proved fruitless. It was impossible to adequately administer legal standards in an industry with a workforce consisting of very small groups dispersed among an enormous number of workplaces.

Eventually the view prevailed that, although some homeworkers undoubtedly benefited from their endeavors, the only way to prevent the exploitation of many more was to ban the practice entirely. Indeed, several major forms of industrial homework were banned more than 40 years ago, utilizing authority contained in the federal Fair Labor Standards Act. All but one of these bans is still in effect (for a fuller discussion of the historical record, see Chamot & Zalusky, 1983).

Note that I am not suggesting that all employers will exploit homeworkers, nor that some people may not prefer to work at home. Today, an extremely small fraction of our workforce is involved with electronic homework, mostly by choice, and we suspect that things will be far worse when large numbers of people are routinely required to work out of their homes. Decent employers will face competition from the less scrupulous sort who, unfortunately, still abound. The AFL-CIO regards the developing situation as so serious that its convention, held in October 1983, passed a resolution calling for a legal ban on electronic homework.

The portability of electronic work presents problems other than for homeworkers. The technology generally frees office work from geographic constraints. It permits the disruption of established work situations at the press of a button. At the extreme, we are faced with export of office work to other countries. This has long been a problem with factory work, and now we face the same possibility for office work. Several examples are already available, such as the move of a large American Airlines data processing operation to Barbados, and the availability of service companies offering routine keyboarding and other office tasks to American firms, using low wage workers in Caribbean countries, India, Mexico, mainland China, and other places. It also is becoming common for multi-national companies, such as automobile manufacturers, to utilize engineers in several countries, linking them and their computers through telecommunications systems.

SOLUTIONS

If we are to avoid further division in our society, and to encourage more cooperation at the workplace, then decisions about future work organization must be reached with the active participation of employee representatives. We must seek approaches that are not only economically sound, but socially viable as well.

Technology optimists argue that economic expansion will create many new jobs, so that we shouldn't worry very much about job losses in the so-called smokestack industries. Growth will come in white-collar and service areas. Yet we see a major possibility for the exporting of office jobs. We can't have it both ways. Just as the establishment of manufacturing subsidiaries in other countries reduced job opportunities for blue-collar workers at home, so, too, will the electronic movement of office work to other countries eliminate potential jobs for clerical workers in the United States.

Thus, the coming of electronic office work brings the potential for serious problems for each group of white-collar workers. Managers and clerical employees, and eventually certain types of professional employees, may face significant declines in demand, and hence, in employment opportunities; some professional employees and many clerical workers will have to deal with deskilling and job rationalization; job-related health problems, including stress, may increase among office workers; and new forms of work organization, including remote sites and foreign locations, will affect many white-collar workers.

None of these problems are inherent in the technologies. Whether they become common or not depend on choices that are made first by engineers and equipment designers, and then by management. There is never only a single engineering solution to any particular technical problem.

The problems we face from office technology revolve around the issue of control. The technologies themselves are neutral. Whether they enhance jobs and make them more interesting, or eliminate jobs or make them more boring and less meaningful, depends entirely on how the technologies are applied. This is much less an engineering decision than it is a question of managerial strategy.

Office technology is moving in the direction of giving managers more control and office workers less control over many aspects of the job. Yet, today's employee wants more involvement, not less. A plan that produces an office of tomorrow strongly resembling the factory of the past does not meet the needs of the employee, nor, I would submit, the best interests of the employer.

Most office technologies are still fairly new, and their full potential is yet to be realized. I have emphasized some serious problems that have already surfaced. Many of the worst problems could be avoided by seeking the active participation of employee representatives early in the change process.

There is real concern for job security, as well as a fear of misuse of technologies on the job. How openly these concerns are confronted, and how well and how fairly they are dealt with, will determine to a great extent the success of the transition to the electronic office of the future.

3 White-Collar Women and the Rationalization of Clerical Work

Priscilla Murolo
Empire State College, State University of New York

ABSTRACT

This chapter examines the rationalization and feminization of labor at the Aetna Life Insurance Company's home office, 1910–1930. Like most large clerical employers, Aetna Life hired women to fill the routine jobs created as skilled work was divided from simple, repetitive operations. Jobs requiring the exercise of considerable judgment in decision making remained the province of men. The gender/skill line was hardened in the late 1920s, when women who held jobs generated by the first wave of rationalization were subjected to a second wave, which did not affect male occupations. The period 1910–1930 was also a time of rapid mechanization of office work. But a study of changing labor systems at Aetna Life during these years shows that it was the managerial drive to minimize labor costs, and not technological innovation, which provided the engine for rationalization and the concomitant creation of a gender divided work force. The chapter concludes by bringing this historical perspective to bear on a brief discussion of the computer's impact on men and women in the insurance industry.

Aetna Life and Casualty's home office is the largest colonial style building in the world. Strangers to Hartford might easily mistake it for the Connecticut State Capitol, and they would not be entirely wrong. For Aetna, the largest diversified financial company based on assets in the United States, is a major seat of power. The power is concentrated in executive suites where company officers preside over a work force that is typical of the insurance industry. The work force is enormous (13,000 in Hartford alone), and its structure is quite complex; but, like all white-collar labor in the monopoly sector, it is divided into two main tiers. The first is the large and overwhelmingly female corps of routine clerical workers whose labor is highly specialized and repetitive, frequently mechanized, and always subject to strict managerial control. The second tier is comprised of the equally vast but mostly male ranks of underwriters, salespeople, and others whose broader knowledge of the business, more diversified jobs, less intensive use of office machines, and relative autonomy within the labor process, distinguish them from the clerical staff.

This bifurcated work force is a product of the early 20th century, when tremendous growth in corporate size sparked a managerial drive to contain burgeoning white-collar payrolls. As the discipline and productivity of office staff became important executive concerns, employers dismantled the traditional labor system under which multifaceted male clerks generally determined their own pace and methods of work and performed whatever mechanical or routine mental operations their jobs required. Tasks that called for the exercise of considerable judgment in decision making were divided from simple, repetitive functions to ensure that high-priced, skilled labor would not be squandered on tasks that could be performed by a cheaper grade of worker.[1] Aetna, like most large clerical employers, hired its first women to fill the newly created routine jobs—at salaries well below the going rate for white-collar men. The line between skilled and routine workers quickly became a gender line. As the female ranks grew, management began to experiment with factory-tested ways of increasing their productivity, the ratio of clerical output to payroll costs. Women who held jobs generated by the first wave of rationalization were subjected to a second wave, which did not for the most part affect men. The two-tiered, gender-divided structure that appeared when skilled and routine labor were divided was solidified by the differential application of managerial pressures for greater output.

The period of rationalization was also a time of rapid mechanization of office work: the mass installation of typewriters, dictating and keypunch equipment, calculators, and so on. Women initially entered the office as machine operators. Employers' investment in new technology assisted in the restructuring of the white-collar labor force, the feminization of routine jobs, and the later speed-up of clerical workers. But it was the corporate quest to minimize labor costs, and not technological innovation, that provided the engine for rationalization and gender divisions. A study of the creation of the two-tiered labor force at Aetna Life makes this abundantly clear and provides a historical perspective key to understanding the impact of today's new technologies on men and women office workers.

---

[1] For a description and analysis of this phase in the rationalization of office work, see Braverman (1974, pp. 315–319).

FROM HOMOGENEITY TO HIERARCHY TO GENDER SEGREGATION

In 1867, when the Aetna Life Insurance Company was 14 years old, it had $91 million of life insurance in force and a total yearly income (including premiums, interest, and dividends) of $5 million; but its home office staff was still quite small and homogeneous. A dozen or so gentlemanly clerks working at stand-up desks were personally supervised by the company's president. By the early 20th century, the clerks numbered 300 (Pierce, 195lb), and the distance between them and Aetna's officers had increased. A supervisory and middle-management staff had developed. Its job was to assign tasks, see that clerks kept busy, and judge the quality of their work. This division of responsibilities signaled the end of the old labor system recalled below by one former clerk.

> Everybody knew what had to be done on the job, and if the work was done that was all that was necessary. If on a fine day somebody wanted to go to a ballgame he would hustle and get all his work done as far as he could go and then ask his desk companion if he would take over if anything unexpected came in. Getting a YES he would then go out the back window into the alley, or more likely through the boiler room . . . and away to the game. Next week it would be another fellow who wanted to go fishing—or something. (George T. Griswold Biography File, cited in Pierce, 195lb, pp. 21–22)

Had a man decided to go to a ballgame after July 1, 1908, his path to the window would likely have been blocked by a division head or senior clerk who, according to the new "Rules and Regulations of the Home Office," was charged with enforcing company policies regarding attendance and hours of work and with making sure that clerks did not talk, smoke, loiter, or make personal phone calls during the business day (Aetna Life Insurance Company, hereafter ALIC et al., 1915). As the labor force grew, rising payroll costs commanded executives' attention, and the productivity of clerks became a serious managerial concern. A disciplinary hierarchy was established to replace the old, informal means of extracting work from the staff.

The first woman to work at Aetna Life was a switchboard operator hired in 1908. (The switchboard had opened 6 years earlier with male operators.) She was the company's lone female employee until 1910, when Aetna decided to compile a Hollerith card record of its data on mortality.[2] Thirty-five women were temporarily hired as card punchers. In 1911, when the

[2] In 1885, Dr. Herman Hollerith devised a machine that could "read" data from cards by sensing where holes had been punched in them. This precursor of the modern computer was used to tabulate the U.S. Census of 1890 and, by 1910, was widely used by large corporations to sort and classify data.

project was completed, a few stayed on (Dixon, 1965). The work force joined by these women was nothing like the autonomous staff of the nineteenth century. But if male clerks were grumbling about the managerial drive for greater work discipline and stricter supervision, it is unlikely that the women heard them. In 1910–1911 the Hollerith card punchers were segregated on the top floor of the Aetna Life building and instructed to use only the back elevators. This arrangement was designed to make certain that they would be invisible to Aetna's president, Morgan Bulkeley, who in the words of one former employee, "approved the [Hollerith] project, but just didn't want to accept women in a business office" (Dixon, 1965).[3]

Five years later Bulkeley did grudgingly accept a permanent female presence. By early 1916, there were 50 women—mostly machine operators—in the home office. By the fall of that year, their number was up to 150. (Pierce, 1951b). They continued to work in exclusively female departments until the United States entered World War I. Then Aetna found it necessary—not to mention cheaper—to replace fighting men with women. By the mid-1910s, Hartford's public high schools had geared up to meet the local insurance industry's demand for routine clerical labor and were offering girls a host of courses in basic office skills. So there was an ample supply of trained women seeking clerical jobs—enough to guarantee that these workers could not command a high wage, even in the midst of wartime labor shortages.

Many women remained at Aetna Life after the war. By 1922 they were 44% of the home office labor force (The Aetna Affiliated Companies, 1923); but they were almost entirely relegated to routine jobs that had become sex-typed as female occupations. Some served as clerical support staff in mostly male departments. The majority worked in all-female environments: typing, filing, stenographic, Ediphone, and card-punching pools. Sexual segregation on the job was built into a labor system in which routine functions were designated "women's work" and most of these functions were centralized. Segregation outside of workrooms was also a company policy. Women were required to use the home office's back entrance, along with separate female elevators and staircases, even after Morgan Bulkeley's death in 1922 (Slimmon, 1925). The gender line at Aetna was reinforced later in the decade, when management launched a campaign to increase the productivity of routine workers.

---

[3] For a discussion of the ideological debate surrounding women as office workers, see Davies (1983, chap. 5).

THE PRODUCTIVITY DRIVE

As one company history has it, 1925 was the year in which Aetna Life began to take "deliberate step . . . to counter the insidious trend to impersonal employee relations" (Pierce, 1951d, p. 29). Yet that same history offers facts that lead unavoidably to another interpretation of management's goals in instituting new personnel policies.

Aetna Life's business was booming. Yearly premium income had reached previously unimaginable highs. The number of salesmen in branch offices had skyrocketed. By 1923, the sales force was so large and so dispersed throughout the country that home office management, worrying about agents' efficiency, began to standardize procedures among them (Pierce, 1951a). Aetna Life officers must also have watched closely as a brother company, Aetna Casualty and Surety, started in 1923 to measure the cost of each operation in all of its 25 branch offices. The results were an insurance executive's dream: Costs fell dramatically, many agencies nearly cutting expenses in half—simply because top management was measuring (Bills, Wallace, Donaldson, & Lies, 1926).

By 1925, Aetna Life had another reason to seek ways of cutting what it regarded as personnel fat. Sales had continued to rise, but insurance-company profits do not depend mainly on policy premiums. Selling policies is a method of accumulating money for investment; millions of small payments by policyholders become finance capital in the hands of company officers and directors. In the mid-1920s, Aetna Life's aggregate rate of return on its investments was beginning to drop. Heavy investments in corn belt farm mortgages, which over a 50-year span had helped the company climb to the top of its industry, were now a serious liability. A widespread agricultural depression was marring Aetna Life's profit picture (Pierce, 1951c). Payroll savings could help to alleviate the pinch.

What the company history does not mention is the fact that in 1925, Aetna Life was more concerned than ever about the number of "slackers" working in the home office. Absenteeism was a problem. Women employees had gotten into the habit of leaving 5 minutes early in order to catch their buses home. Supervisors were apparently unable or unwilling to crack down on the practice (Slimmon, 1925). Top executives had begun to suspect that the day-to-day discipline of their growing work force could not be fully entrusted to lower management, and they were certain that clerical productivity could be raised far beyond present levels. Thus, in the spring of 1925, they instituted a major study of the personnel situation in each home office department (Pierce, 1951d).

Aetna Life was not the only large insurance company to make such a move in the 1920s. As early as the mid-teens industrial engineers and factory efficiency experts had begun to discuss how Frederick Taylor's famous system for the rationalization of industrial work might be applied to clerical operations.[4] William Leffingwell (1917), author of the path-breaking *Scientific Office Management,* reported the effects of his consulting work at the Curtis Publishing Company, where $100,000 was saved on clerical work in 1917 alone (Fuller, 1919). Such results quickly caught the eye of insurance executives. Efficiency consultants from Carnegie Institute of Technology's Life Insurance Sales Research Bureau were soon in demand. And in 1924, the Life Office Management Association, made up of scientific-minded personnel officers from life insurance companies, held its founding convention.

The individual hired to direct Aetna Life's massive personnel study was Dr. Marion A. Bills. In 1917, she had received a Ph.D. in psychology from Bryn Mawr and, after several years of teaching and research, had made her way to Pittsburgh to join Carnegie Tech's Life Insurance Sales Research Bureau. She arrived at Aetna Life in April 1925 and by early 1926 had so distinguished herself in service to the company that she was elected an Assistant Secretary—the first woman officer at any Hartford insurance firm ("Miss Bills celebrates," 1950, p. 16). Barely 36 years old, she was probably one of the youngest officers as well.

A group photograph taken at the 1928 Life Office Management Association (LOMA) convention shows Bills as a serious-faced woman, sensibly dressed in skirt, blouse, sweater, and lace-up shoes. No one could mistake her for one of the many vacationing wives who had accompanied their husbands on the convention trip to Chicago. Bills was obviously there for work to which she was earnestly committed (LOMA, 1928). By all accounts, she was equally business-like at the Aetna Life home office. Combining advocacy of women's rights in the insurance industry with a devotion to scientific office management, Bills' work at Aetna affected women's position at the company in two main ways. She helped to increase the employment of female supervisors and to institute a policy of hiring college women at salaries equal to men's. Her arrival also signaled the launching of a productivity drive in departments where women were concentrated—those in which labor was the most routine, repetitive, and easily measured.

Bills' 1925 report on the home office's personnel situation is not in the Aetna archives, but an educated guess is that it focused especially on the possibilities of measuring individual output in various types of work. Prior to her arrival, Aetna Life management had equated a high level of clerical productivity with a high level of busy-ness. As one set of confidential "Instructions to Division Heads" had explained:

[4] For a first-hand explanation of Taylor's principles and descriptions of some of his work in factories, see Taylor (1911/1967). For a critical discussion of Taylorism, see Braverman (1974, pp. 112–120).

> The work of the Division should be so assigned that the daily task of each employee will keep him steadily at work during the full office hours. Employees absent from their desks for unwarranted periods must be carefully checked. Employees from other divisions must not be allowed to interrupt the work except on strictly business matters. It is only through the strict observance of these practices that waste can be eliminated, and loitering and gossiping in the halls and stairways prevented. (ALIC)[5]

But according to scientific office managers, such practices were of little value. Some believed that it was too easy for employees to appear busy when they were actually shirking. Others, like William Leffingwell, insisted that office workers—"whatever may be the case in factories"—wanted to put in a fair day but were thwarted by lack of training or inefficient management (Fuller, 1919). All agreed, however, that the measurement of individual output was the key to any successful program for eliminating waste. Keeping workers busy was not good enough. They also had to be paid differentially, according to the amount of work they accomplished. Thus they would have a real incentive to work not only steadily but also as swiftly as possible.

In August 1925, while she was studying Aetna Life departments, Bills also oversaw an American Management Association survey of 250 corporate members of the Association's Office Division. A questionnaire asked companies for descriptions of any programs for measuring office output. Only 25 of her nearly 80 respondents both measured and paid some part of their staffs on the basis of that measurement; but payroll savings for companies that had programs were significant. Bills was most impressed with the results obtained by Cheney Brothers, silk manufacturers in South Manchester, Connecticut.[6] Eighty-one of Cheney's 283 clerks received bonuses determined by their measured output; and over the 9 years during which the plan had been in effect, the company had been able to cut its staff in bonus departments by one third (Bills et al., 1926).

By March 1926, Bills was ready to try such a plan at Aetna Life. Her starting place was the Renewal and Accounting Department. Individual output was measured by the number of bills sent out each day. Bonuses brought the average yearly salary of the affected clerks up from $1,067 to $1,507, but gross payroll costs fell as the original staff of 114 was reduced to 66. In addition, the number of items handled by the department increased by 90% (Bills, 1929). The program was, from top management's point of

---

[5] The "Instructions" are filed in the Aetna archives with "Rules and Regulations of the Home Office" for 1908 and 1915, and were probably circulated in the late 1910s or early '20s.

[6] The bonus plan used by Cheney Brothers, and later by Aetna Life, was a variation of the Gantt Task and Bonus System, developed by H. L. Gantt when he was working with Frederick Taylor at the Bethlehem Steel works in Pennsylvania. For scientific managers' evaluation of this scheme and others used in factories, see Hathaway (1923).

view, a resounding success; and Bills was given the go-ahead to install the bonus system in other departments (Bills, 1927).

The general procedure she used was this: First, Bills and her assistants used stopwatches to measure the time absorbed by each of a job's elementary operations. After determining the "one best [i.e., fastest] way" to perform every operation, they fixed what would in the future be the "allowed time" for each unit of output (lines typed, cards punched, records filed, etc.). From that moment on, individual rates of efficiency were figured according to the ratio of "allowed time" to "taken time." A bonus of 1% of the base salary was paid for each percentage point of efficiency over 70%. For example, if a unit of output was timed at 8 minutes and a worker took 6 hours to complete 40 units, her efficiency equation would be:

$$\frac{\text{320 minutes (allowed time for 40 units at 8 minutes per unit)}}{\text{360 minutes (6 hours taken complete 40 units)}} = 89$$

$$\text{89 minus 70} = 19$$

She would receive for that day a bonus equal to 19% of her basic daily salary.

The efficiency equation was used for more than figuring bonuses, however. It also played a key role in reducing staff in bonus departments, for the all-important "taken time" was not reported by workers but ascertained by supervisors who measured each worker's *idle* time and subtracted it from the 390 minutes (210 on Saturdays) of a full work day. Workers were thus given immediate monetary incentive not only to work as fast as they could but also to make the free time, once carefully hidden from management, fully visible. At first the system must have been quite attractive to workers whose department heads had formerly endeavored to keep them "steadily at work during the full office hours." Now workers were invited to bring in books and needlework (Bills, 1929), and they were rewarded with higher bonuses for the time they spent reading, knitting, or sewing. Spurts of extremely rapid work could be followed by long rest periods—until a department reached the next, inevitable stage of bonus installation.

As Bills (1929) described step two:

> When we first put in a bonus, we usually run up about 40 to 50 per cent idle time. That is, the first month. Then we begin to reduce personnel until we get down to probably not more than 10 or 15 per cent [about 45 to 60 minutes per day]. In a service department you should not get much below that because you have to take care of your peak loads. (p. 18)

As had been the case at Cheney Brothers, the number of workers in an Aetna Life bonus department was usually cut by one third. But at Aetna, the cuts were made over the course of months and not years.

There were few lay-offs, for the company's low pay scales (women in routine jobs generally made from $14 to $20 per week) and sanctions against employing married women usually made lay-offs unnecessary. Although her public reports on bonus installation occasionally mention the need to "let go" workers whose efficiency rates consistently fell below 70% (Bills 1927), Bills found that she had "good luck with turnover" in bonus departments (Bills, 1929). The new pace of work and minimization of idle time must have caused some women to quit Aetna Life in search of employers who had not yet concocted productivity drives. But low pay, dead-end jobs, and the availability of work at other local insurance firms combined to give Aetna Life a fairly high rate of turnover in both bonus and non-bonus departments, among lower level men as well as women. Moreover, insofar as female turnover was concerned, a woman's marriage almost always signaled the end of her time at Aetna Life. The overall attrition rate thus made it easy for Bills to reduce staff without direct dismissals. Some workers would leave bonus departments to marry or take different and slower paced jobs with new companies. Others could be transferred out of bonus work and into the vacancies constantly appearing within other Aetna Life departments.

In a few cases, workers who resisted transfer to non-bonus departments were mollified by having their new salaries increased to cover the bonuses to which they had become accustomed (Bills, 1929). It did not cost much to meet the demands of these individuals. Marriage would cut short the employment of most of the transferees, and their newly hired replacements would take over at the usual, lower salaries. On the other hand, once a bonus plan was fully established, protecting the interests of individual workers was of little importance. New women hired into previously bonus-ized jobs seem to have been widely subject to discharge when they could not meet efficiency standards or when an aggregate increase in a department's idle time paved the way for further staff reductions.[7]

In the fall of 1926 Bills, the Renewal and Accouting experience now under her belt, began a gradual campaign to convert all of Aetna Life's routine workers to the bonus system. A year later she was able to report that fully converted departments had reduced personnel by 39% (from 82 to 49 workers) and had lowered their net payroll costs by 25%. The remaining

[7] Aetna archives contain Employment Registration Bureau Records for only one 11-month period: December 29, 1929 to November 27, 1930. During that time, 164 women left the company. Most resigned, usually to marry or to take other jobs. Seventeen left on what appears to have been an involuntary basis. All but 4 of this group had been with the company less than 1 year. The reasons listed for their departures are as follows: 7 were deemed "unsatisfactory." Three left because of the departments' "reduction in force." Of the remaining 3, 1 was "inefficient," 1 judged "not suited to work," and 1 is described simply as "discharged." Bureau records do not list the job titles of these 13 women, but at least 8 of them—from Ediphone, Hollerith, and filing departments—were surely working under the bonus plan when they lost their jobs. See ALIC (1929–1930).

workers enjoyed an average pay increase of 20%—considerably smaller than the 41% increase received by Renewal and Accounting clerks (Bills, 1927). The original experiment had apparently taught Bills how to arrive at an "allowed time" that would keep bonus levels relatively low.[8]

Bills' (1927) detailed description of bonus installation in the Ediphone (dictaphone typing) Division indicates she also had learned that it was not individuals' efficiency but the reduction of both personnel and cost-per-unit by which the company would measure the success of scientific management.

> We established a bonus in this Division on October 1, 1926. For the first month, we showed an average of 37% free time. We then reduced a clerk and in the next two weeks' period, we showed an average of 35% free time. We reduced another clerk and showed an average of 25% free time. We reduced another clerk, but on account of increased efficiency the free time was still 29%. Elimination of a fourth clerk took our average percentage of free time to 24% which was about the free time we felt we should have in a service department such as Ediphone. The number of lines [of typing] have since increased and we have not found it necessary to add to our personnel. . . . That gave us six clerks where we had previously had ten, a reduction in personnel of 40%. The average efficiency of the individuals was then around 90%. We were paying an average bonus at that time of about $4.00 a week, a little less than $20.00 a month, while our net cost per line had gone down from $.0071 to $.0036. (pp. 156–157)

By 1928, Bills had further reduced the average free time in the Ediphone Division to 18%—a fact suggesting that staff and cost reductions continued even after the bonus system was fully in place (LOMA, 1928, p. 175).[9]

Reading between lines of her description, we might surmise that pressure on workers continued as well. Note that when the first clerk left the department, free time fell by only 2 percentage points; but when the second clerk left it fell by 10 points. Now realizing that staff reduction was essential to the bonus scheme, the Ediphone women seem to have been working more slowly, reducing the number of minutes their supervisor could report as idle time, and hoping to avoid further staff reductions. But, when a third officemate was eliminated, the slowdown seems to have been eclipsed by fears on the part of each individual that she would be the next to go. Every

---

[8] In none of Bills' reports on bonus installation is there any mention of workers' contesting the "allowed time" assigned to their tasks. This does not, of course, mean that workers were satisfied with her analysis of their operations; but it does suggest that whatever objections she encountered were not so strong as to interrupt the process of installation.

[9] There is no evidence that the "allowed time" was refigured for any bonus-ized jobs. Further increases in productivity were obtained by staff reductions and the minimization of rest time and not by the recalculation of the figures upon which the speed of work would be measured and bonuses determined.

typist now had to make sure that she was not the least efficient, and therefore the most expendable, of the department's workers. The pace picked up, and free time rose by 4 percentage points. Only when the bonus installers were satisfied that all were working as fast as possible and all free time was fully visible did staff reductions end.

In addition to reducing the cost of a line of Ediphone typing by almost half, Bills had been able to show the workers that their jobs depended on their playing the bonus game by management's rules. Those who did so could increase their earnings from about $20 to $24 per week. Those who could not or would not meet the new demand for speed lost the change for better pay. Given the reality of sub-standard salaries—which for insurance women in nearly every routine job fell below the average salaries available in banks, public utilities, mercantile, and manufacturing firms (U.S. Personnel Classification Board, 1939)—the Ediphone women's informal resistance to the bonus plan could not withstand the onslaught of efficiency experts.

Bills of course, was, delighted with the results of the new system in the Ediphone and other departments. Reporting her triumphs to the 1927 meeting of the Life Office Management Association, Bills (1927) expressed a belief that "probably about one third of our clerical force can successfully be put onto an individual bonus system" (p. 159). At that time, only 134 workers had entered the plan and only 102 survived staff reductions, but Bills was determined to add to their number (Bills, 1927).

Installation, however, continued at a slow pace. By September 1928, the number of workers put onto bonuses had reached only 350, with 250 surviving reductions (LOMA, 1928). Fourteen workers now covered were Ediphone operators, file clerks, typists, Hollerith card punchers, tabulators, renewal checkers, and policy loan calculators (Bills, 1928, p. 4). Fourteen months later, the number entered into the bonus plan had climbed to only 380, and the number of survivors remained at 250. At an American Management Association conference, Bills admitted that she was "a little ashamed" that just one tenth of Aetna Life's 2,500 home-office employees were on the bonus plan. She was still hopeful, though, and predicted that the proportion could eventually reach 60% (Bills, 1929).

THE LIMITS OF RATIONALIZATION

In 1930, Bills stopped giving conference papers on her bonus installation work. There was no further mention of past successes or hopes for the future. Aetna Life seems to have decided that installation was a largely completed process. After the establishment of a central personnel department in 1931, Bills could no longer operate as a free agent in whatever efficiency drives she might conceive. She was now assistant to the chief personnel officer, who had final authority over all efforts to streamline departmental work procedures (Brainard, 1931). The new Personnel Department's initial bulletins to officers and division heads indicate that the company was still primarily reliant not on bonus systems but on the watchful eyes of supervisors to maintain work discipline. Management in all departments—including those with bonus plans—was instructed to crack down on workers who were habitually tardy or absent, who broke for lunch before designated times, or who left their stations for extended periods (Slimmon, 1931a, 1931b).

The Great Depression could hardly have made Aetna Life executives enthusiastic about a productivity drive that involved salary increases. From July 1932 to July 1934 the company, for the first time in nearly 70 years, was unable to pay its stockholders a dividend; and in the fall of 1932 all Aetna employees received a 10% pay cut (Hooker, 1956). Furthermore, a massive installation of the bonus plan during the depression would surely have required layoffs. Hard times had sharply lowered the rate at which Aetna's employees left for jobs with other firms. Young women with needy families and unemployed fiancés were postponing marriage to stay on the job. The company could no longer count on a turnover rate high enough to make bonus-installation layoffs unnecessary. Dismissals of this kind would have been deadly to Aetna Life's carefully cultivated atmosphere of labor peace, in which a tradition and promise of job security were (and still are) essential ingredients.

It would be wrong, however, to conclude that, were it not for the depression, bonus-ization would have proceeded apace. The slow progress of installation in 1928 and 1929 suggests that, as word of speed-ups and staff reductions spread throughout the company, Aetna Life employees were less willing to cooperate with management in the crucial, early stages of bonusizing a department. Furthermore, a great deal of insurance work—particularly that traditionally done by men—did not easily lend itself to the measurement of individual output. Writing policies, evaluating applications, figuring risks: these were not routine or repetitive tasks for which one could easily determine an "allowed time." The men who did this work might follow standard procedures, but they also had to exercise personal discretion in applying general rules to individual problems. Quite a few problems might be so thorny as to require consultation among co-workers. How could individual output be gauged? Applying the bonus system to the bulk of Aetna Life's men would have necessitated measuring the mental labor of employees whom a female back-up staff of typists, machine operators, file and figure clerks had already freed from most routine work. Bills' prediction that 60% of Aetna Life's home office employees could be put onto bonuses indicates that she thought such measurement was possible, at least in some cases. What stopped her from trying it was not any personal doubt about the broad applicability of scientific management to office work but the resistance of supervisors in predominantly male departments.

One of Aetna Life's longstanding personnel policies was to hire outsiders into only the lowest level jobs; lower and middle management was recruited from the ranks (ALIC, n.d.). Thus, when Bills came to Aetna Life, most supervisors were men with seniority in the company. Many had, no doubt, started as clerks in offices where "everybody knew what had to be done on the job, and if the work was done that was all that was necessary." The several executive directives previously cited show that top management had constantly to remind these men of their new responsibilities as disciplinarians over other workers.

Bills, too, believed that tradition-minded men made inadequate supervisors; and, in preparing departments for bonus work, she tried to create an environment in which these individuals could learn new ways. Bills (1929) explained:

> We are convinced that the best management demands that units and divisions be made of sufficient size that a supervisor can devote the major part of his time to supervision. We believe that this is a correct principle for [three] reasons: First, the person doing personal and supervisory work will in nine cases out of ten neglect supervision for personal work and this is true no matter how much you may emphasize the importance of supervision. Second, approaches to the two kinds of work are quite different, and do not as a rule fit into each other. The supervisor of the small group is very likely to be the person who has grown up in the work and has the viewpoint. When the job includes no personal work you eliminate the possibility of this. Third, the mind of the supervisor should be free so that he not only can see that the present work is going through satisfactorily but also can look for improvements in methods of operation, efficiency and cooperative services rendered to other divisions. (p. 12)

As a bonus installer, Bills found it much easier to work with supervisors who had not "grown up in the work."

> While some supervisors of long standing resent any suggestion that the efficiency of their division could be improved by measurement of the work done, practically all of the younger and more progressive ones feel that such measurement gives them a chance to show what can be done and an assurance that when it is done, it will be appreciated (Bills, 1929, p. 13).

Appreciated, though, by whom? The resentful old-timers Bills mentions were, it seems, men who had decided to forego top management's pat on the back in order to safeguard morale within their departments and to maintain their camaraderie with workers they supervised. Without a bonus system, they could be flexible in overseeing a department's work. They could, for example, let employees linger over lunch and feel certain that the favor would be repaid with a faster work pace if backlogs began to appear. The bonus scheme, which promised to "personalize" employee relations by paying every individual according to his or her output would in fact have disturbed the mutually beneficial personal ties that had traditionally bound workers and supervisors together in what many no doubt viewed as a common departmental enterprise.

The attitude of tradition-minded supervisors explains, in part, why bonus installation ceased once it had covered those newer departments whose work forces were almost entirely female. Not only was women's work the company's most routine, repetitive and measurable, but their supervisors were also the least likely to have come up through a female department's ranks. Rationalization schemes were more agreeable to those who had never done the same work as their underlings and who, consequently, had fewer personal ties and loyalties to the workers they supervised. Add this to the fact that male supervisors were less likely to take seriously the preferred work rhythms of young women in routine jobs and you had in Aetna Life's typing, filing and machine-operating departments a scientific manager's playground.

Not so in older departments, where women worked as auxiliaries to men and where supervisors had long experience as "one of the boys." For example, when Bills investigated the possibility of applying the bonus system to life underwriters, she hit a brick wall. As she remembered it: "With six section heads, I sat through an entire afternoon trying to find out whether [the underwriters'] job was routine or not. We were totally unsuccessful. Every man went away with the same opinion as to whether the job he had come to discuss was routine or not" (Bills et al., 1926, p. 6). Resisting the bonus plan was relatively simple for these men. All they had to do was insist that the work they supervised was too complex and required too much critical thought for measured output to reflect labor input. And who was Bills—a woman and an Aetna Life novice—to disagree?

Only the most aggressive support from top management could have enabled her to introduce bonuses into departments like Life Underwriting, and that Bills did not have. Instead, in the mid-1930s, the company adopted a system that might well be viewed as a compromise between upper management and supervisors who resisted becoming either strict disciplinarians or the measurers of idle time and solvers of efficiency equations: Productivity experts would not enter departments with stopwatches or bonus plans. (Bills, confined to Personnel, had been put to work on the development of clerical aptitude tests to be used in hiring and placement). At the end of every year, each department's production performance would be evaluated on the basis of personnel cost per unit of output. Internal work arrangements were left up to supervisors, who were pressured by the Personnel Department only insofar as requests for additional staff might not be met if unit costs were deemed too high (Pierce, 1951d).

Bills thus remained "a little ashamed" of the proportion of bonus-ized jobs at Aetna Life. Although the proportion of bonus workers grew along with the amount of repetitive and measurable work, it never came close to her goal of 60%. Today, most of Aetna Life and Casualty's typists, file clerks, photocopiers, keypunch operators, coders, form-completion clerks, and mail-room workers earn efficiency bonuses. The grand total receiving bonuses is now about 2,000; but that is only 15% of the company's Hartford work force (M. Rodriquez, Productivity Services Department, Aetna Life and Casualty, Hartford, CT, personal communication, April 22, 1982).

Trained as they were in the more complex aspects of life insurance work, it is doubtful that the supervisors of men in skilled and diversified jobs ever seriously feared losing to the bonus drive. Indeed, years before the unit-cost compromise had been reached, the drive had become somewhat of a joke among men whose jobs could not be fully rationalized. The November 1931 issue of the sale force's *Life Aetna-izer* (p. 24), for example, printed the following:

Office Manager—"I'm afraid you are ignoring our efficiency system, Jones."

Jones—"Perhaps so, sir, but somebody had got to get the work done."

CONCLUSION: RATIONALIZATION AND THE TWO WHITE COLLARS

Although many of the details are peculiar to Aetna Life, the early history of women at this company exemplifies the rationalization of office work in the entire United States insurance industry. Other giant firms, such as Metropolitan Life, preceded Aetna in the reorganizing of the traditional labor process and in the feminization of routine jobs. But all firms eventually divided skilled from routine work; and in every case, the vast majority of the specialized and repetitive jobs became the province of women. Productivity drives designed according to the principles of scientific management also were widespread. Proceedings from the Life Office Management Association conference in the 1920s show that bonus plans, the standardization of work methods, and other schemes for speeding up white-collar labor were numerous and that women—stenographers, typists, machine operators, low-level clerks— were always the central target. It was not only at Aetna that feminization and rationalization went hand in hand.[10]

[10] A great many large clerical employers used scientific management to rationalize office work in the 1920s (Davies, 1983, pp. 97-128). The movement seems to have been especially vigorous within the insurance industry, but it touched many other types of firms as well. Men in routine jobs—shipping clerks and office boys, for example—were frequently the targets of rationalization, but women, because they outnumbered men in the lower grades of office work, were especially affected wherever scientific management was applied.

Women's presence in the industry diminished the effects of rationalization on men. Two distinct white-collar labor forces were created. One was made up almost wholly of women, whose routine work was carefully designed and paced by management. At Aetna, this group was concentrated in, but by no means confined to, bonus departments.[11] The other force, which included a relative handful of women, escaped the productivity drives of the 1920s and thus retained much of its customary control over the pace and methods of work. Skilled clerks, underwriters, and others within this group were subject to increased discipline as a supervisory and managerial hierarchy developed; but they were spared the chief effects of rationalization. When much of insurance work was deskilled, they held on to comparatively skilled jobs. The difficulty of measuring their individual output, combined with their supervisors' resistance to scientific management, later protected these workers from systematic speed-up. Their autonomy within the labor process remained far greater than that available within feminized occupations.[12]

The labor force in the insurance industry is still overwhelmingly segregated by gender. The installation of automated systems is changing its occupational structure, but gender divisions remain intact. Between 1961 and 1980, a period of rapid computerization, life insurance occupations became even more sex-divided (Feldberg & Glenn, 1983). Women are still concentrated in the most routine jobs, although the titles and content of these jobs have changed with computerization.

Gender divisions disguise much of the labor rationalization to which computer technology has been harnessed. Underwriting is a case in point. It is the only life insurance occupation that is becoming more sex-integrated, and it has been largely deskilled by the computerization of risk evaluation for standard policies (de Kadt, 1979). The job has been and is being degraded, but it represents a step up for women. In other cases, software that encapsulates the knowledge of skilled men is used in the creation of entirely new jobs assigned to women. These jobs frequently offer higher pay, more diversity, and more autonomy than the occupations previously open to women. Female labor under the new set-up is not, however, as well paid or autonomous as was skilled male labor within the old system (Feldberg & Glenn, 1983). The feminization associated with computer-assisted deskilling pro-

---

[11] Women holding routine jobs in mostly male departments, such as Life Underwriting, did not fare so differently from the Ediphone operators. Though they were surrounded by men and a few relatively privileged women whose individual output could not be scientifically gauged, their own output was easily measured by informal, if not scientific, means. Their jobs did not prepare them for upgrading into skilled insurance work. The fact that they could never be seen as "one of the boys" and the specialized, repetitive nature of their tasks meant that they could be subject to output pressure from even the most tradition-minded supervisors.

[12] The one female occupation which remained relatively untouched by scientific management was that of private secretary (Davies, 1983, pp. 128–139, 168–169). Insurance companies, unlike many other corporate bureaucracies, have never employed a large number of secretaries.

vides some small opportunities for women; but it also closes off the possibility of women's entry into the higher level jobs being removed from the labor process. The only unalloyed advance is made by employers in their movement to minimize labor costs.

Along with deskilling, the enduring corporate drive for greater output motivates computerization. On this front, too, the current wave of rationalization has a differential effect on men and women. The aggregate idle time of both groups is reduced when jobs are combined under automated systems that redesign and reassign work. But it is primarily women, lower level clericals at computer keyboards, who are subject to the harshest pressures for increased productivity. Computers—which can count key strokes, claims adjusted, forms processed, etc.—make it easier and cheaper than ever before to measure individual output; and routine workers' weekly pay is often determined by this measurement.

Although it affects them differently, computer-assisted rationalization is aimed at insurance workers on both sides of the gender/skill line. In the early 20th century, the separation of skilled from routine labor and the feminization of lower level jobs created a situation in which women in the 1920s bore the brunt of the drive for greater productivity. Employers' current use of the computer to deskill labor—putting skills into software—makes men, too, the objects of rationalization. Women are still on the front lines insofar as speed-up is concerned. Whether men will escape the measurement of their output, as they escaped the bonus system at Aetna and other companies, is an open question. Sexual segregation of occupations and male privilege within the bureaucracy still characterize the insurance industry; but rationalization within automated systems is changing the working conditions of both sections of the two-tiered labor force. The corporate quest for a greater ratio of output to payroll costs is the main factor determining the computers' effects on both men and women.

ACKNOWLEDGMENTS

This chapter is based on a paper prepared for a seminar in 20th-century American labor history, conducted by Professor David Montgomery of Yale University.

I gratefully acknowledge the assistance of Leith Johnson, archivist for Aetna Life and Casualty, Hartford, CT. Constructive criticism of the first draft was provided by members of the seminar: Dana Frank, Ken Galdston, Jack Gillette, Toni Gilpin, Dan Letwin, Lizzie Reis, and Jonathan Solovy. Special thanks to David Montgomery for his comments, suggestions, and encouragement.

4 Changing Office Technologies and Transformations of Clerical Jobs: A Historical Perspective

Suzanne Iacono
Public Policy Research Organization
University of California, Irvine
Rob Kling
Department of Information and Computer Science
University of California, Irvine

ABSTRACT

Office life and clerical jobs have dramatically changed in the last 100 years. In the last decade, there has been a substantial move to computer-based office systems, and clerks play a significant role in their use.

Some advocates of the office of the future suggest that clerical jobs will qualitatively improve after integrated multifunctional computer-based office systems are used. Other analysts emphasize that computerization has reduced the quality of some clerical jobs. In contrast, we argue that computer-based systems can support better jobs or worse jobs.

Despite some changes in the character of worklife and the division of labor, computer-based office systems do not automatically improve or transform clerical jobs by opening new career lines or by making clerical workers less interchangeable or more valued. During the last 100 years, clerical jobs have developed as a lower and relatively isolated sex-typed stratum of white-collar work. Significant changes in clerical jobs require organizations to make choices that include allowing their clerks to influence the organization of their jobs, increasing investments in developing their organizational skills, and valuing their work.

Clerical tasks—writing and transcription of written symbols—accompanied the first writing systems many millennia ago. Clerical jobs—the bundling of writing, copying, and various forms of record keeping—were first held by clerks and priests. By 1512, a clerk was "a subordinate employed to make written entries, keep accounts, etc." (Onions, 1955). The term *clerk* derived from *clerical,* which connoted the literate priests who copied religious manuscripts. Through the 19th century, clerical jobs of all kinds—bookkeeping and secretaries—were held by men and were relatively privileged positions, although despised by craftsmen and manual workers (Delgado, 1979). Clerical tasks are almost synonymous with office work, but clerical jobs are social constructions in which bundles of clerical tasks are delegated to particular workers.

The character of office life and clerical jobs has dramatically changed since 1512 and even in the last 100 years. Offices have become cleaner and brighter; more large offices with finely specialized jobs have emerged. Massive office buildings housing thousands of white-collar workers, often in different organizations, dominate the urban landscape. Many managers and professionals perform routine clerical tasks, but often denigrate them as "mere paperwork." Mere paperwork is the substance of clerical jobs, even when records are kept in computer files rather than on paper forms.

Clerical jobs that were once held by men apprenticing themselves for a profession or a management position now are virtually feminized and segregated from careers in management and the professions. Clerical jobs were a tiny fraction of the workforce in 1900, but now they constitute about one fifth of the available jobs.

Alterations in the character of office life and clerical jobs are not solely derived from changes in workplace technologies although many analysts focus on new technologies as the central causal agents in determining social outcomes (Giuliano, 1982; Nora & Minc, 1980; Strassman, 1985). Other factors such as changes in market conditions, the available labor force, decision-making processes, and the power structure in society also play major roles (Pape, 1983). However, as Kraut (this volume) indicates, the history of work is closely related to changing workplace technologies. In offices, the technologies have changed from ink wells and quill pens to electromechanical machines such as telephones, photocopiers, and computer terminals. New technologies do not simply replace old technologies. However, their introduction can be the occasion for restructuring work and changing office practices. In the last decade, there has been a substantial move to computer-based office technologies and clerks play a significant role in using them.

The technologies of office work have dramatically improved over the last 100 years and are likely to improve during the next decades. Do improved technologies for office work—those that are faster, quieter, more powerful, easier to use, and cleaner—lead to better office jobs? In this chapter we examine one aspect of this question by focusing on the role of computer-based technologies in altering clerical jobs, especially secretarial jobs.

COMPUTERIZATION AS A COMPLEX STRATEGIC INTERVENTION

It is a mistake to conceptualize the computer as a single technology because different computer system architectures offer different opportunities for structuring work. Through the early 1980s, computing in offices has included two major forms: computer-based information systems and word processing (WP) systems. Clerks often enter and retrieve transaction level data from information systems. The majority of WP operators are clerks.[1] In the mid-1980s office computer systems run on stand-alone micros and on a new generation of relatively integrated equipment that support specialized information systems and the flexible manipulation of text: electronic mail and schedules, as well as WP. Although these new systems are aimed at managers and professionals, many clerks use them to support the work of managers in their offices. For example, some private secretaries may serve as gate-keepers for incoming electronic mail. Some secretaries also routinely download and upload data from a mainframe and organize data for office reports.

Many analysts suggest that the use of these multifunctional computer-based technologies will radically transform the worklives of office workers, including clerks. Accounts of these transformations focus on "offices of the future" where each desktop has a micro-computer or terminal on it. Enthusiasts argue that this new type of office will differ significantly from past offices due to the flexible nature and extensive use of new computing technologies (Giuliano, 1982; Poppel, 1982). Many analysts who write about an office of the future assume that improved information technologies will also improve working conditions for office workers (Giuliano, 1982; Poppel, 1982; Strassman, 1985; Uhlig, Farber & Bair, 1979). Jobs will be more interesting, creative, and less routine. People will communicate more easily

[1] In a recent survey of computer-based text processing, Rittenhouse & Kling (1984a) found that about one-third of word processing machine users are non-clerical: professional staff, supervisors, and managers.

and have more data at their fingertips. As a result, clients will be better served and offices will be more productive.

Other analysts emphasize the negative impacts of computing technologies on clerks. They argue that managers and professionals will monopolize benefits in the form of increased control and enhanced productivity while deskilling and regimenting clerical jobs (Bjørn-Anderson, 1983; Gregory & Nussbaum, 1982). Baran and Teegarden (1984) argue that office automation in insurance companies alters work in a somewhat more complex pattern: data entry work will be reduced; some professional work will be automated and passed on to clerks whose jobs will appear upgraded; and the remaining professional work will be more specialized.

These two main lines of analysis rest on economic and technological determinisms: The character of work is determined by the nature of the technology and the structure of economic interests. We disagree that the character of work can be deduced a priori from the abstract characterization of technological capabilities. But we do not believe that technologies are merely incidental to the transformations of work (see Murolo, this volume, for a contrasting perspective).

Another alternative perspective, views the introduction or alteration of computer-based systems as *complex strategic interventions* in organizations and their offices (Kling, 1983, 1984; Kling & Iacono, in press-b). Organizations make many kinds of social and technical choices about computer technologies and the ways in which support and access to them are organized. Alterations in the character of worklife and the division of labor are the consequences of social and technical strategies that accompany the introduction of technologies rather than as necessary or predetermined consequences.

This strategic interventionist view is important because it encourages analysts and participants to take account of the variety of software and hardware technologies that support computer-based office systems and the variety of work arrangements that can accompany their implementation. Some mixes of computing technologies and work arrangements can be beneficial for many or most participants; other mixes of computing equipment and work organization can lead to poor working conditions for some or many workers. Still other mixes can result simultaneously in some advantages and some disadvantages for any one stratum of white-collar jobs. (We provide a detailed example later when we examine the forms of work organization that support computer-based text processing.)

Computerization has improved some jobs and diminished the quality of others (Attewell & Rule, 1984; Kling, 1980b). They are influenced by the particular strategic interventions and the social and technical constraints and opportunities of the larger organizational setting. An organization with abundant resources has more potential for making choices that will enrich the working environment than those that are operating within a tight budget. For example, most staff need training to master new technologies and to feel comfortable using them. When technologies are introduced without adequate training, many workers will experience new demands and uncertainty instead of new opportunities. When staff who are not especially fascinated with new computers are forced to work with "better" computer equipment but receive little training, they may use it less well and feel more stress than when using technologically more primitive equipment that they understand.

Deterministic models attract attention because they present simple and compelling stories about concrete causes that have straightforward consequences and they can be described in lean and clean prose. For example, if new computer-based applications are more flexible, then the working conditions of those who use them will be more flexible; if systems are more rigid then jobs will be more rigid. In practice, however, these links are less direct. Deterministic models portray simple worlds of social actions: Single causes can be isolated, and they have single effects. Our strategic interventionist perspective portrays a world of social action that is more complex and more uncertain—causes are harder to isolate and they can interact to produce complex outcomes. No single statement about worklife in an office of the future is likely to be accurate for all mixes of technology, work organization, and the detailed practices that accompany new equipment.

The technologies and social character of offices past and present are an essential base on which to interpret and understand worklife in offices of the future. This chapter examines the relationship between changing office technologies and the character of clerical jobs. First, we examine five transformations in the quality and character of office life over the past 100 years. Then we examine transformations in clerical jobs during that time period and implications for the future. Will working conditions and career opportunities for clerks improve with the use of new office technologies? If so, under what conditions? Last we examine the possible character of clerical jobs in offices of the future.

TRANSFORMATIONS IN THE QUALITY AND CHARACTER OF OFFICE LIFE

During the last 100 years, millions of offices have served as workplaces in the United States. Offices today differ substantially from those of 1880 as a product of many local changes in equipment, the organization of work, and practices about hiring, firing, promotion, and pay. If strategic interventions in these matters are aligned in many offices, they can add up to major trends in the larger economy. The first offices to hire women clerks, establish typing pools, install telephones, purchase systems furniture, or adopt microcomputers were unusual (Delgado, 1979). When tens of thousands of offices do these same things, they restructure the worlds of office work and labor markets for a substantial fraction of the office workforce.

The strategic interventionist perspective has been best developed by organizational designers who ask what should be done regardless of local traditions or even larger historical trends. The primary unit of analysis is the office or organization. Even though it is usually viewed as an open system, it also is viewed as manipulable as a separate social unit. In contrast, historical sociologists summarize millions of interventions taking place over decades in a few simple patterns (e.g., Mills, 1951; Ritzer, 1977). In this chapter, we try to appreciate the options for changing clerical jobs within organizations by using the interventionist perspective. Simultaneously, we try to understand how clerical jobs have developed over the last century and what larger institutional patterns have developed around them that may make some changes easy and others difficult or unlikely. We know of no good systematic data for comparing office life over the past 100 years, but the literature about offices suggests five major transformations in the quality and character of office life:

1. *Physical working conditions in offices have generally improved.* Most offices today look and sound different from their pre-World War II precursors. Since 1930, high-ceilinged rooms with incandescent lights, heavy desks, and fixed, solid walls have been replaced by rooms with fluorescent lights, systems furniture, and movable pastel partitions. Most larger office buildings are centrally heated in winter and air conditioned in the summer. While some buildings are drafty or chilly in winter and quite warm and stuffy in the summers, we suspect that the general comfort level of most offices has improved in the last 100 years. Overall, the new technologies usually are more convenient and effective, and fewer people work in dingy offices. However, newer offices are not utopian habitats: some are noisy, stuffy, highly enclosed, lack speech and visual privacy, etc. (Makower, 1981).

2. *Office work has become increasingly dependent upon a diverse array of complex technologies.* Historically, the direction, coordination, and execution of office activities have been supported by technological "devices of the day." In 1800, office equipment consisted of pens, ink, and proper paper. At the turn of the century letter-press and very primitive typewriters expedited clerical tasks (Whalen, 1983). Punched-card tabulating equipment spread in accounting offices, and telephones spread widely in the United States between the two world wars (Whalen, 1983). Since World War II, we have seen office technologies change dramatically from manual typewriters, carbon paper, ballpoint pens, and check-writing machines to WP, photocopiers, felt-tip pens, and general purpose computers. Electronic media have become commonplace.

3. *The economic scale of office work has increased.* Most white-collar workers work in offices. There is no systematic data about long-term changes in the number of office workers, although one can use employment in selected occupations such as general management, law, or clerical jobs as

rough surrogates. Office workers grew from a tiny fraction of our agrarian economy in 1880 to about 40% of the 1982 workforce of 100 million (Panko, 1984).[2] The Department of Labor classified about 18.5 million as clerks in 1982. Of about 30 clerical categories reported by the BLS, about 3.8 million (or 21%) were counted as secretaries. Taken together, bookkeepers, cashiers, office machine operators, and typists counted for another 5.7 million jobs.

4. *Very large offices with highly specialized work organizations have emerged.* In the 20th century, we have witnessed the rise of mammoth organizations with proportionately large office support—large insurance companies, banks, engineering design firms, and government bureaus. The social boundaries and physical boundaries of an office are diffuse. In a few organizations, some offices are vast rooms with hundreds of staff, separated by low partitions and systems furniture. More typically, organizations build hard walls to contain work groups of fewer than 50 people, whether the groups are called a "section," a "department," a "group," or a "division." These larger office organizations set a context for the activities of smaller work groups: the larger office organizations often work with large volumes of data that can justify extensive automation. They also may have more subtle and complex procedures for handling exceptions than smaller organizations. Technological and social innovations take longer to implement in larger organizations. Last, these larger organizations often employ armies of clerks and are a major site of employment for clerks.

5. *The proportion of women working in offices has increased tremendously.* Women began to enter offices in the 1890s with the introduction of typewriters. The proportion of women in the workforce has increased steadily since the 1800s (although there were short-term peaks, such as during World War II). But women are concentrated in several sex-typed occupations: sales clerks, school teachers, nurses, and clerical jobs (Matthaei, 1982). About 35% of the 42 million working women are clerks (U.S. Bureau of the Census, 1983). Kraut (this volume) asserts that new office technologies have most directly affected clerks, and thus they have changed the job content for women much more than for men.

---

[2] Despite the widespread use of "office" as a term which seems to denote a kind of workplace, there is no clear definition of office in most of the sociological or office automation literature. The differences between casual and analytical uses is illustrated by calling the place that a dentist drills teeth an "office." Acceptable definitions are crucial for estimating the size of an office workforce (Panko, 1984:212). Panko excludes dentists and even postal clerks from his estimate of office workers, but includes computer equipment operators. While some of these choices appear odd, they do not alter the main point, that office work has grown substantially in scale. They do, however, influence one's estimates of the actual scale of office work in the economy and the actual kinds of work that "officer workers" do.

Even though office life has been transformed along these five dimensions, offices today vary considerably. For example, the work performed in the Oval Office, an insurance claims data entry office, a doctor's office, and an engineering design office all differ significantly. Offices differ along several key social dimensions: size and number of employees; physical arrangements; work organization and diversity of tasks; predominant control strategies; degree of formalization; and division of labor.

The technologies of office life also vary considerably. Office automation can be viewed as complex strategic interventions that allow many social and technical choices. These choices include the kinds of hardware and software that are acquired; the distribution of equipment throughout the organization; practices for training staff in new technologies; and the ways in which work is organized or reorganized, for example. Sometimes the computing environment develops slowly over time, with many different kinds of equipment used in various departments throughout the organization. In other situations, there may be large scale interventions at particular points in time.

Offices also can be characterized in information processing terms: They are workplaces where the manipulation and communication of data and information are the primary tasks. Some formal information processing activities in offices include transcribing, copying, filing, planning, supervising, deciding, and communicating.

These various tasks are executed through a division of white-collar labor. The Bureau of Labor Statistics classifies white-collar work into five categories: managerial, professional, technical, sales, and clerical. Within this division of labor, a hierarchy of roles identifies who has authority to implement specific decisions and set schedules and who abides by them. There are many choices in ways to divide work, but clerical jobs are more likely to be at the bottom of an office's social order than are other white-collar office jobs.

Clerical jobs often are composed of supervised support tasks closely tied to office machines and office procedures. Although many professionals and managers routinely engage in some forms of clerical tasks (writing prescriptions by hand or writing memos or reports on PCs, for example), their major work activities derive directly from their professional or managerial positions. These white-collar workers often have more discretion in deciding which clerical tasks to delegate and which they will do themselves. Clerical jobs, on the other hand, consist entirely of delegated clerical tasks that have remained "low end" work for the last 100 years. During this time, clerical jobs have changed in five major ways:

1. The fraction of clerical jobs in the United States economy has increased.
2. Clerical jobs have become feminized.

3. Clerical jobs have become segregated from advancement to professional or management jobs.
4. The technical support of clerical tasks has improved.
5. Working conditions have not improved for clerks as much as for other white-collar workers.

We discuss these transformations in detail throughout the rest of the chapter. In the next section, we analyze some important historical transformations in office life and their effects on clerical jobs.

TECHNOLOGICAL CHANGE AND CLERICAL JOBS

For the past 100 years, clerical jobs have been the lowest stratum of white-collar work. Even when clerical tasks improve, clerical occupations still remain lower stratum jobs. The technologies and forms of office life have changed remarkably since 1880. Dozens of technologies that make some tasks easier have been introduced into offices. Despite these enhancements, office staff still are expected to work hard.

We view the deployment or alteration of computer-based technologies as a complex strategic intervention. Outcomes vary according to the choices that powerful actors make about structuring or restructuring work in their organizations, even though some consequences are unanticipated. Some clerical jobs have been enriched because they require more skills and are more interesting (Kling, 1980a). Other clerical jobs have lost some specific work tasks that were redistributed to other more specialized departments in their organization (Murphree, 1984). But regardless of changes in job content, clerical jobs within the larger occupational structure have not been transformed in pay, influence, or status.

The white-collar hierarchy is based on a division of labor that has institutionalized clerical jobs into one occupational stratum with common features: low pay, delegated work, and blocked career lines (Kanter, 1977; Matthaei, 1982; Ritzer, 1977). Some organizations have adopted computer-based systems and reduced clerical staff while reintegrating some clerical tasks into managerial and professional jobs. But we know of no organizations that have systematically reintegrated managerial and professional activities into clerical jobs.

We now examine some of the dimensions that have structured clerical jobs over the past 100 years and will continue to be important as office technologies change in the near future.

Status and the Segregation of White-Collar Jobs

Clerical jobs were predominantly held by men in 1880. For many, a clerkship was an entry point to learn a given business and develop a career within it by working up from the bottom. Professional schools were weak or undeveloped outside of medicine, theology, and law. As a consequence, entry into most professions, especially commercial ones, came through training on the job. The development of typewriters in the late 1800s helped create secretarial jobs (Scott, 1982). Men who had previously taken jobs as clerks began to enter administrative and professional positions through other gateways. By 1900, the segregation of male and female office work had begun: New secretaries were women and administrators remained male.

In the early 20th century, most firms sought to hire young women aged 18-25 into secretarial jobs (Scott, 1982). Because married women stayed at home, these jobs were usually organized on the assumption that secretaries would leave work after marriage by their mid-20s. For example, most of the insurance companies located in the Hartford, Connecticut area had, between 1910 and 1930, explicit policies of hiring only unmarried women (Murolo, this volume). These jobs could be dead-end by design and not be part of a social world in which women "learned the business" because they were temporary jobs. Explicit policies of hiring unmarried women as clerks and maintaining low pay were complex strategic interventions carried out by insurance companies during that era.

Moreover, entry routes into various professions increasingly depended upon some formal post-high school education. Although there was substantial variation across occupations, and even within them, clerical apprenticeships no longer provided opportunities for moving into positions of increased status, authority, and decision-making.

On-the-job training for most clerks today is not extensive given a base of knowledge about general office procedures and skills on some office machines. Compared with other white-collar workers who receive more organizational investment in terms of training and learning the business, clerks are more interchangeable and locked into a structure that prevents them from developing the particular skills that could result in career lines outside the stratum of clerical jobs.

The segregation of the labor force into a relatively cheap and interchangeable pool of clerks and a more highly trained group of managers and professionals is one managerial strategy for reducing costs. Since the early 1900s, the bifurcation of white-collar work has been gender-based. Today, with changing attitudes and opportunities for women, managerial and professional jobs are desegregating. However, clerical jobs remain solidly female (Burris & Wharton, 1982). In 1950, 62% of all clerks were women. By 1970, 74% of clerks were women and today, 80% are women (Gregory & Nussbaum, 1982). Conversely, clerical jobs are a significant form of employment for women. About 35% of the 42 million working women are clerks (U. S. Bureau of the Census, 1983).

The white-collar hierarchy is based on the power and authority invested in managers, not only on skill (Mills, 1951). Most white-collar workers are well educated whether they are male or female. In the early part of this century, office work was suitable for middle-class young women who had some formal education and who wanted to do respectable, clean work (Scott, 1982). Bikson and Gutek (1983a) studied offices in California with multifunctional computer systems and found that only 16% of all the office workers had not attended college. Although many technical and professional positions require specialized education, general college preparation is often adequate in many white-collar occupations. Gutek (in press) argues that women clerical workers often are better educated than male blue-collar workers, but are paid substantially less than men with lower educational achievements.

Clerks support activities for all levels of an organization. For example, they file, type, transfer documents, or communicate messages at the request of their superiors. They may be unaware of the consequences of letters they have typed or messages they have relayed. Typically, they do not influence key organizational decisions nor the substantive issues of their own work.

Ability alone does not explain the sexual segregation of office work. A recent national survey of managers and secretaries carried out by the Honeywell Corporation (1983) reported that many secretaries feel they are capable of taking on more responsibilities than managers say they are. Strassman (1985) argues that secretaries will necessarily take on greater responsibilities in highly automated offices because they can leverage overall productivity by using a wide array of their skills. Although we agree that this is a valuable idea, we do not believe that saying so will make it so. Moreover, secretaries constitute about 21% of clerks. Many clerks, such as telephone reservationists and data entry clerks, are leveraged into highly productive work organization through routinizing work rather than creating enlarged jobs.

During the last decade, the skills in manipulating machines required for general secretarial work have been increasing. In 1970, most secretaries were required to have typing skills. Some temporary placement agencies now prefer secretaries to be skilled on several machines. These new skills, in operating more complex photocopiers and WP, do not appear to open new career lines for secretaries. They primarily indicate the ways in which the technologies of office life have become more complex.

Clerks most often are rewarded for finishing segmented clerical tasks on schedule. It is more important for clerks to complete delegated tasks such as typing documents than understanding what they mean or how they effect their organization. Understanding the work flows in the larger organization is not considered part of their job. There are no incentives for keeping abreast of the latest developments, and there is little encouragement to search out information, join professional organizations, or attend conferences.

In a recent survey of text processing supervisors, 70% of the respondents reported that there was little or no opportunity for advancement for text processing workers into the larger organization (Rittenhouse & Kling,

1984a). Less than half of the respondents had heard of professional organizations such as the International WP Association (now the Association of Information Systems Professionals), the Data Processing Management Association or the Association for Computing Machinery. Of the supervisors who had heard of any of these organizations, less than 60% reported ongoing involvement in them.

The introduction of new or enhanced computer-based technologies has not radically changed social relations in most offices. Clerical jobs remain low status jobs even when more skills are required for operating more complex machinery. The segregation of clerical jobs from other white-collar jobs is a strategy for maintaining an occupational stratum of low paid and dispensable workers. Most office clerks are women. As a result, women can expect to benefit least from using new "improved" technologies in terms of increased status, pay, and career opportunities (Iacono & Kling, 1984).

Working Conditions

Working conditions in offices have generally improved over the last 100 years. Most physical aspects such as temperature, lighting, and furniture are adequate. Some organizations take special pride in pleasant interior environments and modern architecture. However, when work space and other social amenities are distributed throughout the office hierarchy, clerks receive relatively less than other white-collar workers. This is as true today in modern office settings as it was in the factory-like buildings of World War II. Clerks are the most likely to suffer from a lack of privacy or physical segregation from others. Often their individual work spaces are relatively small, cramped, and cluttered. The low status and segregation of clerks are reinforced by the conditions of their physical environment.

Machines are an important part of any office. They often are located in or near clerical work areas. For economies of scale, expensive equipment is often located in public areas where managers can easily see the equipment being used. But more importantly, clerks are expected to be the most intensive users of office equipment. Most of their work is tied to one or more machines. Other white-collar workers have more discretion in their use of equipment—what they will use and when they will use it. Even if they have workstations in their offices, they are not expected to constantly use them. In the extreme case, full-time data entry clerks, travel reservationists, telephone operators, and customer service staff are tied to computer-based data systems and sometimes telephone headsets. They often have limited physical mobility and their time, as well as productivity, is usually closely monitored. Regimented work organizations defined by machine use shapes their worklives. Some of these telephone-respondent jobs are shaped more by telephone technologies than by computer technologies, even when they are used conjointly. They are more telephone-mediated than computer-mediated. (See Zuboff, 1982, for a contrasting interpretation.)

Managerial concern with the productivity of new office equipment has primarily focused on clerical usage, even though the integrated office systems also support professional and managerial users. However, even when they have been successful in achieving the desired levels of productivity, other problems may result. For example, Turner (1984) found that the use of an "improved" on-line system compared with an older teletype system used by clerks in the Social Security Administration resulted in increased productivity but decreased job satisfaction. The clerks interacted with more clients each day because the improved system was faster and decreased the length of each client interview. Increased client contact meant more emotional interchanges and dealing with more client problems that led to increased absenteeism, mental strain symptoms, and an overall decrease in job satisfaction.

Clerks are more frequently dissatisfied with the working conditions that accompany new technologies than they are with the technologies themselves. Enid Mumford (1982) illustrates how the causal relations between altered technologies and social outcomes are usually complex.

A department in a large British firm purchased two stand alone word processors for the use of a group of five secretaries. Space was limited, and the machines were placed in an isolated room. The secretaries had been used to typing reports and letters with occasional interruptions from telephone calls and other office staff. They could work efficiently without interruption when they wanted to use the word processors. But, they hated the new arrangement. Now, they had to work at a continuous tempo moving their eyes from "paper to screen," for an hour or two without significant changes of tempo and few breaks in relative isolation. The secretaries generally liked the flexibility of the new equipment, but they strongly disliked the new working conditions under which they used it.

In one sense, the WP systems were the catalysts (and thus "cause") of a changed work arrangement that led the secretaries to work under pressure, in relative isolation. The information processing capabilities of WP were not the causal agents. Although the information processing capabilities made them attractive to the director, their cost (about $18,000 apiece) led him to insist that they be used efficiently. They were not sufficiently compact to replace a standard typewriter and they required that some additional space be found. The combination of space shortages in the department, the director's desire to have secretaries use expensive machines "efficiently," and a social order that places secretaries at the bottom, led them to work under pressure in isolation.

In some cases, workplace technologies have altered jobs. For example, telephone operators' jobs have changed considerably since the 1940s when long distance operators had to know routing patterns to direct calls across the country. The jobs always tied operators to headphones and consoles. But as telephone switching was automated, the operators required less skill and the jobs became more tedious. But the large majority of clerical jobs have not been so unambiguously altered with technological changes.

The vast majority of individual secretaries who work for higher level managers are satisfied with the productivity of their automated equipment (Honeywell Corpation, 1983). In a study of computing in the worklife of traffic-ticket clerks, Kling (1980a, 1980b) found that most traffic-ticket clerks attributed positive, job enlarging characteristics, such as increased variety, to their use of automated equipment. However, Mumford's case illustrates how clerks can be highly pleased with new technologies and simultaneously frustrated with the conditions under which they use them. Moreover, working conditions can vary with different types of arrangements. Next, we examine three types of work organization for computer-based text processing.

Work Organization for Text Processing

Strategies for organizing text processing work vary. In an on-going study of computer-based text processing (Rittenhouse & Kling, 1984a, 1984b), we found that three major approaches were common: centralized arrangements, decentralized arrangements, and mixed arrangements. Each type of work organization represents a complex strategic intervention for organizing access to computer-based text processing equipment. In the extremes, some of these work arrangements are highly regimented, whereas others permit substantial flexibility in workers' access to equipment and their control over time. But more commonly, the clerks view these different work arrangements as having both advantages and disadvantages and do not find any one preferable in every way.

A WP center can give employees several advantages not provided by a decentralized arrangement, such as a longer career path. For example, in one organization a woman started as a secretary and moved up to executive secretary, normally the top of the ladder when clerical jobs are decentralized (Rittenhouse, 1985). When the company brought in a new WP system, she actively developed a WP center, and soon was named its supervisor. Later she was promoted to manager of administrative services, where she directed all of the company's secretarial functions.

These career opportunities were available because a WP center is a complete support organization with its own hierarchical structure. For example, private secretaries often depend on their immediate bosses for advancement: Manager and secretary often move up an organizational hierarchy as a team.

Clerks in a WP center have a second advantage: They develop a special expertise with their systems by working with them constantly. Because of economies of scale, they tend to have the best equipment in an organization.

A centralized WP organization has disadvantages as well. The work is starkly routine and can be dull. There is less flexibility in determining the order of tasks. The social contact with workers and clients outside the WP center is limited; operators become segregated, in touch primarily with a supervisor who collects work from various departments and passes it along. The issue of monitoring job performance also arises more often than in a decentralized work organization.

On the other hand, clerks in decentralized office arrangements, with individual managers having personal secretaries or several professionals sharing one secretary, have considerable flexibility in determining their work schedules and the order of tasks. The job can be more interesting than in a centralized arrangement, because they have more contact with coworkers outside the clerical job stratum, and strict monitoring of job performance is less common.

Decentralized clerks also have fewer opportunities for advancement and do not acquire the same level of technical expertise as workers in a WP center. From a political perspective, the individual secretary is also at a disadvantage. The clerical organization is fragmented, with secretaries having stronger ties to their bosses than to each other; therefore collective job action, such as forming a union, is more difficult.

Some organizations have both individual secretaries or local arrangements and a WP center. The advantages of such a mixed arrangement are obvious for managers: They have local support for immediate needs, and they can lean on the WP center for larger projects that would be difficult for one person to handle, but could be completed in a day if divided among several workers. For the clerical staff, the mixed arrangement also provides more options, but tension can result from the division of clerks into two types.

In one such organization the WP workers saw themselves as much more professional than the traditional secretaries because of their higher level of technical expertise (Rittenhouse, 1985). The secretaries did not have the opportunity to become intimately familiar with the WP system because their jobs were more varied. Also secretaries were sometimes asked to do menial tasks, such as get coffee for the boss, whereas no one in a WP center was asked to perform personal menial services. In addition, some secretaries resented the WP specialists because they were more likely to use the best equipment.

These variations in the kinds of jobs that secretaries and WP operators have with WP mirror variations in the character of clerical work. Some clerks, particularly bookkeepers and general secretaries, have significant discretion over the tempo of their work and related working conditions. At the other extreme, people who perform one narrow set of tasks on one machine, such as fielding telephone inquires with an on-line data base, or full-time data entry, sit in one place for long periods of time, perform rather repetitive tasks, and have their work closely monitored.

In most cases, it is not the technology that makes these jobs bland and stressful, but the way that work is tightly organized to extract the most work from machines and their clerical operators. Clerks are expected to be the most intensive users of office machinery. When information technologies can be used with some discretion, and are relatively reliable, clerks can find them helpful. For example, access to a better text formatter may reduce the amount of difficult formatting problems and consequently make some tasks easier and less time consuming to perform. But new computer-based text processing equipment may also make some discretionary jobs worse when the equipment is unreliable or needlessly complex, when training is inadequate, or work is more pressured to make the equipment productive. Information technologies can be the occasion for restructuring work (Giuliano, 1982; Kling & Iacono, 1984; Markus, 1984).

Whether jobs are better or worse after a computer system is installed or altered depends on the particular bundle of social and technical interventions. Critical concerns for clerks include the level of skills required, job performance monitoring, pay, job security, career lines (Grandjean & Taylor, 1980), and social relations of the restructured activities. Although these may be improved with some strategies of computerization, these strategies are far from universal. For example, Bjørn-Anderson (1983) reports that it has been common for systems analysts to develop simplified clerical applications that are "fool-proof" and to deskill non-secretarial clerical jobs in the process. He also argues that the polarization of office jobs into more interesting professional jobs and more regimented clerical jobs often follows Taylorist strategies for using computer systems to enhance office productivity. (See Murolo, this volume for a detailed description of Taylorization in clerical work.) New equipment costs money. Unless the equipment vastly speeds up some clerical tasks, there is some attraction to reducing overall costs through further regimenting jobs or skimping on pay.

In summary, most clerks have little discretion in the pacing of their work, decisions about which technologies they will use, and the arrangements in which they will use them. The low status and segregation of clerks are reinforced by the conditions of their physical environment. They are more often dissatisfied with the working conditions that accompany the introduction of new technologies than with the technologies themselves (Form & McMillan, 1983).

The complex strategic interventions that have accompanied the deployment of computer-based text processing in organizations have resulted in three major types of work arrangements: centralized, decentralized, and mixed. Each of these three arrangements has both advantages and disadvantages. The centralized arrangement appears to have more opportunities for career advancement than the other arrangements because of its own hierarchical structure. However, the career ladder is short. In addition, WP supervisors do not believe that opportunities exist for WP clerks to move out into the larger organization. Will the next generation of computer-based technologies dramatically alter the quality and character of clerical tasks? In the next section, we examine alterations in the character of worklife and the division of labor in future offices.

CLERICAL JOBS IN FUTURE OFFICES

In the next decade, managers will purchase a new array of computer-based office systems. Many of them will be more multifunctional than current systems, and will support text processing, electronic mail, calendars, schedulers, laser printers, and local area networks, as well as specialized databases. The equipment will typically be more powerful and at its best will further ease clerical tasks. Will clerical jobs also improve in these offices?

Like the offices of today, the offices of the next decade will vary substantially in the kinds of equipment they use. Some organizations will invest heavily in sophisticated technologies, whereas others will try to save money by using older or simpler machines, and make them less available. Similarly, organizations will develop highly varied working arrangements for clerks. Alterations in the character of worklife and the division of labor will depend on the specific intervention that organizations make when they adopt or alter their office computer systems.

Much of the public discourse about the office of the future is based on marketing scenarios. Offices are pictured as clean, well organized, neatly furnished, paperless, sunny, full of plants, and each spacious desktop has a micro-computer or terminal on it. These pictures portray a steady, efficient, and a somewhat relaxed and cooperative mode of office work. Secretaries who look like models (actually models dressed as secretaries) complete the picture with their happy and eager faces. Once new technologies are placed in offices, we do not expect more sunlight, new furniture, more plants and everyone to be dressed fashionably. Nor do we expect office life to be more serene automatically. Zones of conflict that exist today will remain in offices of the future. Conflicts over pay, prerequisites, workloads, deadlines, division of labor, and the handling of flexible time, for example, will still exist. Although some new equipment may help make some people more productive, decisions on who is going to work on which task, on what schedule, and how hard will still have to be made. Without changes in the organizational hierarchy and structure of decision-making, clerical workers can still expect to have less discretion about the organization of their work than other white-collar workers (Kling, 1983).

Alterations in the Character of Worklife

The promise of new technologies to improve organizations and worklife dominates much of the public discourse about office automation. Even when evidence about these promises is inconclusive and contradictory, they can provide managers with rationales for the implementation of state-of-the-art equipment (Iacono & Kling, 1985; Kling & Iacono, in press-a). Clerks have the least influence in determining the pace of introducing new office technologies and the ways in which work is restructured and yet they are expected to be among the most intensive users of many new technologies. This is especially true for clerks in full-time data entry jobs, especially insurance claims processors, telephone operators, travel reservationists, etc. As a result, we expect that clerks will be strongly affected by the implementation of enhanced computer-based systems.

However, we expect a wide variety of outcomes that depend on the particular strategies of computerization. Managers are not completely free in selecting the strategies they prefer. Some strategies will be more feasible under certain organizational conditions, such as organizing larger work groups or having more resources. For example, in large organizations with many routine clerical activities, one can expect large regiments of clerks to process highly routinized transactions. But one might also find more flexibility in the implementation and use of computer-based systems. Because organizations have many complex choices, each with a bundle of outcomes, it is difficult to predict a single future office unless the vast majority of offices automate identically.

Few studies link work outcomes to careful descriptions of the changing technologies and social organization of work. Recent studies by Murphree (1984), Rittenhouse and Kling (1984a, 1984b) and Turner (1984) find that clerks do not necessarily find better jobs in offices with the most intensive investments of state-of-the-art equipment.

New computer-based systems don't automatically provide clerks with opportunities for increasing involvement or learning the business. In fact, managerial concern for efficient use of expensive state-of-the-art equipment may result in more stringent performance monitoring for some clerical computer users. This practice virtually eliminates any time that could be available for clerks to learn more about their own work and their organization.

The social organization of offices that helps regulate the conditions of work is more important than the person–machine interface. High quality interfaces can reduce tedium of a single task, but they don't alter the mix of larger tasks, workloads, or other elements of worklife. Scenarios of future offices often portray managers and professionals with desktop computer access. Usually they work in closed or more private office settings than clerks enjoy. Simply because a machine is present does not imply that they will keyboard all day. Rather they are expected to use their computers for more non-routine activities: accessing critical data when necessary or creating better ways to do things, for example. It is possible to focus on new computational capabilities and ask questions like, "how can we best use 1 Mips machines on every desk?" Questions framed in this way do not help us understand important differences between clerical and other white-collar workers' use of new office technologies.

The adoption or expansion of computer-based technologies does not necessarily have to result in worse jobs. Organizations with resources may increase investments in training clerks for office-specific technical configurations and combinations of technologies so that they can give other office staff more technical support.[3] Then some clerks can become specialists in the use of the technologies and become more valued in their local environments.

Porcella (1985) found that WP staff located in local end-user settings were less interchangeable than those staff working in larger centers that served the entire organization. Presumably the local WP staff came to learn the detailed procedures and techniques of their own office technologies and consequently other white-collar workers in their office came to depend on their knowledge and skills. However, managers in organizations with the small local WP centers disliked the fact that their clerical staff were indispensable and suggested that they would like to change that in the future. It appears that for clerks to become more valued and less interchangeable managers will have to value clerical work and clerks. Increased investments in clerical training and work arrangements that facilitate the acquisition of new skills are not enough to insure that clerks will be more valued in the future.

In organizations with few resources, training may be inadequate. The adoption of new technologies may place more demands on clerical workers who are already overloaded with work. New equipment requires time to learn and space in which to locate it. For example, the addition of more machines into cramped clerical work areas usually makes them more congested.

Some analysts have indicated that the flexibility of the new technologies will be mirrored by more flexible and, thus, "better" working arrangements (Giuliano, 1982). In some cases, organizations have set up satellite work centers closer to employees homes (Nilles, Carlson, Gray, & Hanneman, 1976; Olson, 1983). In other cases, workers can telecommute from home or while traveling. Olson and Primps (1984) found that work at home increased stress for women with low status clerical jobs and primary child care responsibilities. In addition, they were socially isolated, had reduced work status, lower compensation, and fewer benefits. Only one woman in the sample had a professional position. All of the other women occupied clerical positions and believed that the arrangement was their only way of providing an income and taking care of their children. Few clerks have found that

[3] Examples of these new kinds of activities include knowledge about how to transfer data from one system to another, knowledge of the protocols of several systems, the ability to use text editors on several machines and to reformat documents from one text editor to another.

more flexible working arrangements have translated into a less stressful worklife at home as suggested by the more enthusiastic advocates of new computer-based technologies.

Some strategic interventions that clerks may prefer might include: (a) the organization of work into small local WP centers where they can specialize in the techniques, procedures and business of their local office and still have more career opportunities than in a completely decentralized arrangement; (b) increased organizational investments in training them to use new equipment and in understanding the substance of their work and that of the larger organization; (c) incentives and available time for expanding their skills and knowledge; and (d) commensurate pay for these increased skills.

Interventions like these will have to be implemented in thousands of organizations, to alter the institutionalization of clerical jobs as the lowest stratum of white-collar work. In the short run, increased pay for more skills may be feasible but we don't expect that clerical salaries will rapidly double or match those of professionals. Even if some clerks become specialists in using their new technologies, there is no indication that these new skills will open up career lines into the larger organization. There are large financial incentives in having lower paid clerical specialists available to support the work of higher paid managers and professionals.

Alterations in the Division of Labor

Within the organizational hierarchy, various kinds of office work always have been delegated to clerical staff. Strategies of computerization may change the kinds of work that are delegated. Some kinds of work may be delegated outside the work group or to specialists. Or previously delegated work may be reintegrated into professional or managerial white-collar work. But we do not expect that delegated work will disappear.

Some universities have given microcomputers to faculty with the understanding that they will subsequently reduce their requests for secretarial support. The intention is to reduce employment costs by reintegrating some previously delegated clerical tasks back into the worklife of faculty and hiring fewer secretaries in the near future. Depending on the amount of work delegated to clerical staff in the future, the work of the remaining secretaries may change. It is possible that they may act more like receptionists if the faculty chose to do most of their own keyboarding, editing, and printing. On the other hand, it is inconceivable that faculty will have so much free time that it will not be necessary to delegate some of those tasks. Smaller tasks such as letters and memos may be easy to delegate. Faculty also may desire that the secretaries have compatible equipment so that they can take over some of the more routine aspects of their larger writing projects such as entering required changes from editors or reformatting bibliographies. Secretaries may be required to learn new systems, new keyboard functions, new text editors, and other software such as spelling checkers. As such they will be able to support the work of the faculty. Because demands on faculty time are typically great, the second scenario seems more possible than the first one.

Other complex strategic interventions delegate some clerical work to other groups outside a specific department. At one Wall Street legal firm, legal secretaries lost some of their work to new specialized and automated departments (Murphree, 1984). The secretaries previously helped lawyers research cases, and this work was delegated to a department of paralegals. In addition, some of their more routine clerical tasks were turned over to WP centers and editing departments. As a result, their remaining work consisted mainly of non-routine activities that had to be expedited through the slower bureaucratic procedures and the day-to-day emotional support of their managers.

The content of delegated work often changes with the deployment of new technologies, but delegated work won't disappear. Some new clerical tasks in computerized offices include archiving files, reformatting documents, altering data in prearranged formats, producing computer listings, printing letters, etc. Since clerical jobs consist of support activities at low pay, it is economically rational for higher paid managers and professionals to delegate some of their clerical work, even if they do some of their own keyboarding.

Some clerical job categories may be eliminated with increased use of new office technologies. The BLS tallies approximately 50 categories of clerical occupations. Silvestri, Lukasiewicz, and Einstein (1983) project decreases in some office clerical occupations between 1985 and 1995: data entry operators, stenographers, central telephone operators, and reservation agents and transportation ticket clerks. Other clerical jobs with less than average projected growth include hand bookkeepers, court clerks, file clerks, typists, and shipping and receiving clerks. However, openings for receptionists and secretaries are expected to grow more than average. Overall, they expect that the number of clerks will increase through 1995. But due to office automation, the growth will be average rather than faster than average as it has been for the past 2 decades.

Some clerical jobs also have been created through computerization. The number of computer and peripheral-equipment operators has increased about 314% from 126,000 to 522,000 between 1972 and 1980. Between 1972 and 1980, clerks increased from 14.2 million to 18.1 million (U. S. Bureau of the Census, 1983). These 4 million jobs grew primarily through general economic expansion rather than through new job opportunities provided by new office technologies.

Because the current technologies have not resolved managerial concern with office productivity, the promise of new technologies will still be framed in the language of increased office productivity and decreased costs. One humanizing element for the office of the future might include the reintegration of clerical tasks into the mainstream of office work. This implies a restructuring of white-collar occupations. This involves a risk, however, because many clerical jobs that are now primarily held by women might be eliminated.

It will take more than new technologies to improve the character of clerical jobs. Clerical jobs in offices that adopt these new office technologies will not be significantly less insular or more valued without explicit collateral changes in work arrangements, managerial attitudes about clerical work, and increased organizational investment in clerical workers. The complex strategic interventions that accompany the deployment of computer-based technologies are critical components in determining the kinds of social outcomes that will emerge. The treatment and organization of clerical jobs embody many social choices. These are not "givens" any more than the technological choices for office work are "given."

CONCLUSIONS

The most profound conclusion for a chapter like this would be to provide an accurate portrait of clerical jobs 20 or 30 years from now. Simple projections are most feasible when one uses a simple model of social change, such as a technological or economic determinism. We have explained how computerization in organizations and offices is a strategic intervention in which there are many social and technical choices. The character of clerical jobs 20 or 30 years from now will be the product of hundreds of thousands of such interventions that are not easily predicted. They are easiest to predict if the vast majority of organizations behave identically and if they computerize with substantially similar strategies. Although there may be some common patterns of computerization—through influences that range from changing labor markets to new fads in office organization—there is a great deal of potential variety.

We have emphasized an historical perspective for several reasons. First, one appreciates the range of possible social choices in organizing clerical jobs by learning about the variety of previous arrangements. Second, one can see that clerical jobs—as an occupational stratum—have not changed substantially during the last 30 years despite the adoption of a rather elaborate set of office technologies—from photocopiers through basic computer systems. This leads us to be skeptical that new technologies alone, even exciting ones, will be a strong catalyst to improve the worklives of the vast majority of clerks.

Office of the future scenarios typically exaggerate the role of new and interesting technologies because they often package together other, unrelated improvements. For example, the physical conditions of future offices are held to improve along with the computerized equipment with the added implication that all office workers will benefit equally from these improvements (Giuliano, 1982). Moreover, by portraying a workplace free of continual minor conflicts over pay, status, workloads, deadlines, these scenarios distort the social relations of office life of the present and likely future.

Despite the dramatic improvements in office technologies over the past 100 years, career opportunities and working conditions for clerks have not similarly improved. Although clerical tasks today require more skills in using a more complex array of technologies, these skills are not reflected in increased status or pay (Baran & Teegarden, 1984; Bikson & Gutek, 1983). A new generation of integrated computer-based office systems will not automatically alter the pay, status, and careers of clerks without explicit special attention.

Computer-based technologies neither automatically degrade and deskill clerical jobs nor improve them. The squads of clerks who work at terminals 8 hours a day in specialized data entry jobs may have some of the worst jobs; private secretaries who have discretionary use of multi-functional workstations in the course of varied work tasks may have more interesting jobs. These differences are due more to the ways in which work is organized than the technologies in use.

But women, who dominate the clerical workforce, have a large stake in new office automation because they have little discretion in the pace, scale, or manner by which these technologies are deployed. Most of the social outcomes that can be attributed to the complex strategic interventions which accompany office automation are alterations in clerical job content. Some jobs are enriched; they are more challenging or interesting. Other jobs are worse due to increased monitoring or de-skilling. Other jobs have mixed outcomes. But regardless of these changes in job content, clerical jobs within the larger occupational structure have not been transformed. Clerical jobs remain at the bottom of the office social order. This sphere of work is effectively feminized and segregated from work that leads to advancement in the organizational hierarchy.

Decision-makers, motivated by the promise of new technologies, may care more about office productivity and cost-savings than the impacts of technological change on clerical staff. The elimination of some clerical job categories may be one result. Other impacts may include the necessity for increased skills in the manipulation of complex technologies and increased demands on clerical staff to support their efficient and productive use. If significant improvements are to be made in clerical jobs, clerks must influence decisions about how their work will be reorganized with new technologies (Kling, 1983, 1984; Mumford, 1982).

ACKNOWLEDMENTS

We would like to acknowledge Janet F. Asteroff, Kathleen Gregory, Robert G. Rittenhouse, and Alladi Venkatesh, for their helpful comments on an earlier draft of this manuscript, and Robert Kraut, who provided some especially helpful editorial guidance. This research was supported under NSF Grant 81-17719.

5 Technology and the Transformation of Clerical Work

Roslyn L. Feldberg
Massachusetts Nurses Association

Evelyn Nakano Glenn
Florida State University

ABSTRACT

This chapter analyses the impact of technology on the organization of work and on employment prospects from both a historical and a feminist perspective. The body of the chapter was written in 1979—five years before the conference that gave rise to this book. The authors chose to present this work in its original form followed by an addendum. The original piece offers an "historic" perspective on a contemporary debate, emphasizing both the sexist and especially the managerial bias in earlier research. The addendum discusses how the debate is being addressed at present and the questions still outstanding. In particular, it indicates how the general increase in feminist scholarship has led to new concerns, concerns that are reflected throughout this book. The authors see the shifts and continuities in the debate as highlighting the way transformations of clerical work create new opportunities for and new constraints on the employment of women from varied races, ages, and educational backgrounds.

With the introduction of a new type of automated technology into the post-World War II office, a debate began about the probable impacts of that technology on the organization of work, and on prospects for employment. Large issues were debated: the quality of work, the transformation of the occupational structure, the efficiency and productivity of the office, and the extent of employment/displacement among office workers. On each issue, opposing claims were advanced. Taken together, these claims comprise two positions: the exuberant technologists and the skeptical critics.

Exuberant technologists viewed the new technology as a liberating force that would free the office and the larger society from long-standing problems.

Menial jobs would be eliminated; new jobs, relying on a better educated, more autonomous work force, would be created. These new jobs would derive in part from the work required to manipulate and maintain the new technology. Increased efficiency and productivity in the office would lead to lower costs and allow businesses to provide more and better services. Impacts on employment were seen as non-problematic. New employment would be created as services expanded or, at the very least, attrition would take care of job reductions, because the work force was largely composed of those who had high rates of turnover. Little was said about the prospects of clerical workers who remained in the offices or those who would seek employment there in the future. Apparent problems were dismissed as "short run"; in the long run, the new technology would prove to be a boon.

Skeptical critics viewed the new technology less optimistically. They saw new problems being created in employment. Many of the new jobs associated with the new technology were characterized as even more routine, narrow, and repetitious than the jobs they had replaced. They questioned whether the automated system increased autonomy and efficiency or whether it was primarily increasing management's control of the workers and the work process. They drew attention to the fact that many of the new jobs were not only "fragmented," but also "dead ends" in the occupational hierarchy. Some pointed out that these jobs were assigned largely to women. Finally, some critics argued that as automation expanded, more and more workers would be displaced and there would be fewer jobs, particularly at the middle levels, for the next generation of office workers.

Two features of this debate are remarkable. First, it is still being conducted, and largely from the same two positions. Second, despite 25 years of debate, we will still lack systematic information about office automation and its impacts.

As we view the debate, much of the disagreement arises from a failure to separate out different levels of the social structure that technology can affect. We can identify three different levels: the societal level that is evident in changes in distribution within the overall occupational structure and the differential growth of different industrial sectors; the organizational level that appears in changes in the organization of a firm and its internal occupational and authority structure; and the work process level, which can be viewed in terms of the work flow within work units and the activities of individual workers. Although these levels are related, they are not, therefore, isomorphic. When the levels are not differentiated, analysts leap to conclusions about one level on the basis of evidence from another level. For example, researchers might look at the overall occupational distribution and find that the number of technical jobs within the clerical occupations has increased greatly. They might then conclude that jobs have been upgraded as a result of technological innovation. What they have done is to jump levels. If they looked at the level of the work unit or individual worker, they might see that most jobs, other than the technical ones, have become narrower, less autonomous, and more rationalized.

In this chapter, we examine the relationship between technology and the transformation of clerical work by looking separately at each of the three levels of social structure. Most studies have focused on the societal or organizational level. Our primary interest is on the work process level, especially as changes at this level affect women clerical workers. At each level, we piece together available data in a way that elucidates implications for the workers and the work process. We begin by examining the introduction of computers into offices and their relationship to shifts in the occupational structure. Next, we summarize some case study findings on changes in the occupational structure at the organizational level. Then, we look at the rationalization of the work process in particular units, emphasizing actual work activities. In the conclusions, we discuss the limitations of this literature arising from both sexist and managerial bias. On the basis of available information, we argue that recent changes in clerical work continue trends established earlier: rationalization, transfer of decisions from female clerks to new male occupations, and increasing feminization of clerical occupations.

THE INTRODUCTION AND SPREAD OF COMPUTER SYSTEMS, 1950–1978

Office automation was generally defined as the application of electronic data processing to the work of the office (i.e., "the use of machines capable of recording and storing information and of converting data to whatever uses the organization may have for them. . . . (Hoos, 1961, p. 6). The history of this development is largely impressionistic—a collection of specific instances and general estimates of the use of computers. The installation of the first UNIVAC computer in the Bureau of the Census occurred in 1951, the first use of a UNIVAC for payroll processing at General Electric in 1954 (Greenbaum, 1979). A 1957 survey of 177 industrial firms indicated that 80% had already moved toward mechanization of their office work and others expected to move in that direction shortly (Hoos, 1961). Estimates reported in 1958 were of 2,000 computers in operation, and an equal number on order. The exact meaning of these figures is not clear. Some firms used more than one computer, others replaced older models with newer ones.

Despite the absence of exact measures, the general outlines are clear. In the early 1950s computers were first harnessed to office work. By the end of the decade many large organizations were using computers as part of their routine activities. A 1963 survey of the largest companies in the insurance industry offers a detailed look at trends in the early period. Only two of the companies reported using computers in 1954. Each had one computer. Five years later, 107 of the companies had computers, with an average of 1.7 computers per company. By 1963, 226 companies had or had scheduled installation of computers, with an average of 3.0 computers per company (U.S. Dept. of Labor, pp. 10–11).

The trend continued, accelerating over time, so that computers are now used in the offices of virtually all large and many smaller businesses, and in every industry. A 1976 report on data processing spending lists every major industry, from oil production to apparel manufacturing. Moreover, this picture is consistent with daily experience. In dealing with almost any organization, one finds forms designed to provide information to computers.

As computers became more widespread, the uses to which they were applied increased. The particular kinds of uses, of course, reflect the business of the industry and the size of the company, but the general pattern of uses can be seen by looking again at the insurance industry. Initially, the operations which were computerized were complex numerical tasks such as actuarial analysis, mortality studies, and valuation of reserves. These had never been the tasks of the clerical staff and so had relatively little effect on the organization of their work, although it added new groups of workers to the office (keypunchers at the lower level, computer operators, programmers, and analysts at the higher levels). Next, computers were applied to the large volume, standardized calculations that had previously involved large numbers of clerks; such as payroll or premium billing. These tasks had earlier been carefully specified and subdivided into a series of operations. This rationalized organization of work made conversion to computerized processing relatively quick. Now the clerks in the offices were more directly affected and the focus of work was shifting from carrying out the tasks to preparing information so that the computer could process it. The next stage was the application of computer capabilities to centralized record keeping. Where each department of the insurance company had previously kept its own records, a computer system was developed that could store all information pertaining to all aspects of the account. This was the stage of development in the mid-1960's.

The developments of the next 15 years centered around improvements in the flexibility of the system. New forms of data entry were marketed that eliminated keypunch cards by permitting direct entry via magnetic discs. Terminals (Cathode ray terminals) with electronic screens made it possible for clerical workers to retrieve information from or enter new information into the computer right at their desks. By 1976, it was estimated that over 1.5 million terminals were in use (Datamation, 1977). Centralized data banks linked home offices to outlying branches by telephone. Automatic correspondence systems, which automatically sent letters to all accounts displaying certain codes, were in the experimental stage; while word processing systems were becoming routine in many large companies.

OCCUPATIONAL STRUCTURE

We turn first to changes in the occupational structure that occurred since the introduction and spread of automation. The assessment of impacts of office technology on the occupational structure is complicated by other economic and organizational changes that have occurred simultaneously with growing automation. Some of these developments have had as great, if not greater, effects on the size and distribution of the clerical work force; in many cases, these developments have had impacts opposite to those created by automation. Thus, although technology may have reduced the number of workers needed to carry out a given volume of a particular activity, increased business and shifts in the economy increased the volume of activity more rapidly than the achievement of savings in labor time. Therefore, many more workers were needed.

During the period of expanding automation, there was a general expansion of clerical occupations as the volume of record keeping and communication grew. The growth in these activities can be attributed to increased size and complexity of organizations and the disproportionate expansion of the service and financial sectors of the economy (see Glenn & Feldberg, 1977). During the period from 1960 to 1978, employment in clerical and kindred occupations doubled. Table 5.1 shows the numbers employed in major office clerical categories and in computer specialties for 1960, 1970, and 1978. We can identify from that table the categories that grew more quickly than average, i.e., more than doubled, and those that expanded more slowly, i.e, increased less than two-fold. In some cases, the divergence from the average can be attributed to office technology.

The impacts of automation can be most directly seen in the growth of occupations involved in preparing data, setting up and programming systems and processing information and the decrease in those working in occupations whose activities have been taken over by computer processing. The most dramatic changes have been in the growth of computer equipment operators and computer specialists (programmers and systems analysts). Computer operators increased 200-fold between 1960 and 1978, and more than tripled just between 1970 and 1978, when nearly 400,000 workers were operating computer and peripheral equipment. Computer programmers and systems analysts, who are listed under "Professional and Technical Workers" in the census grew 28-fold between 1960 and 1973, from a tiny group numbering fewer than 12,000 to one numbering nearly half a million. About three-quarters of them were men. During the same period, the number of non-computer machine operators fell drastically. Tabulating and calculating machine operators who numbered 60,000 in 1960 had disappeared completely by 1978 as numerical operations were taken over by computers. Stenographers, once a major category, fell to less than one-third of their 1960 total

TABLE 5.1

Employed Persons in Selected Office Occupations by Sex 1960–1978[a]

Selected Occupations	1960 Men	1960 Women	1960 Total	1970 Men	1970 Women	1970 Total	1978 Men	1978 Women	1978 Total[b]
Computer Specialists	8,528	765	9,293	204,945	49,945	254,601	329,132	98,868	428,000
Programmers	5,970	2,170	8,140	124,991	36,391	161,382	181,051	65,949	247,000
Systems Analysts	2,558	772	3,280	68,293	11,748	80,041	119,928	32,072	152,000
Bank Tellers	39,725	93,205	132,930	34,439	215,037	249,476	38,165	410,835	449,000
Billing Clerks	8,039	35,091	43,130	18,838	87,347	106,185	19,992	148,008	168,000
Bookkeepers	153,559	773,980	927,539	227,214	1,259,443	1,536,657	170,190	1,659,810	1,830,000
Cashiers	112,275	373,955	486,230	136,954	695,142	832,096	180,987	1,222,013	1,403,000
Social Welfare Clerical Assist.	—	—	—	261	942	1,203	—	—	—
Clerical Supervisor	28,393	25,824	54,217	64,391	48,389	112,780	75,072	128,928	204,000
Collectors bl. & act.	26,137	6,485	32,622	32,947	18,537	51,484	32,994	45,006	78,000
Counter clks. except food	47,483	74,156	121,639	76,584	152,667	229,251	85,956	291,044	377,000
File clerks	23,426	121,591	145,017	65,221	294,678	359,899	39,039	233,961	273,000
Messengers	49,659	8,766	58,425	46,206	11,225	57,431	62,031	24,969	87,000
Office Machine Oper.	80,706	230,488	311,194	145,477	408,001	553,478	213,366	613,634	827,000
Bookkeeping/billing	5,432	45,951	51,383	6,720	56,629	63,349	5,985	39,015	45,000
Calculating	671	36,406	37,077	3,071	31,880	34,951	—	—	—

Computer/peripheral equipment	671	1,257	1,928	83,072	34,241	117,313	163,881	229,119	393,000
Duplicating	8,066	5,650	13,716	8,788	11,609	20,397	—	—	—
Keypunch	43,691	117,376	161,067	27,964	244,957	272,921	12,012	260,988	273,000
Tabulating	18,143	7,530	25,673	4,128	4,042	8,170	—	—	—
Office mach. noc.	4,032	16,318	20,350	11,734	24,643	36,377	—	—	—
Payroll, timekpg.	43,920	63,681	107,601	48,622	107,364	155,986	59,045	181,955	241,000
Receptionists	11,018	145,709	156,727	16,046	288,326	304,372	18,228	569,772	588,000
Secretaries	42,607	1,424,166	1,466,773	64,608	2,640,740	2,705,348	28,720	3,561,280	3,590,000
Statistical Clks.	59,042	78,124	137,166	89,204	160,492	249,696	90,103	286,897	377,000
Stenographers	11,625	258,554	270,179	88,097	120,026	128,123	9,024	84,976	94,000
Typists	25,468	496,735	522,203	57,272	922,804	980,076	35,496	1,008,504	1,044,000
Misc./not specified	448,521	1,085,922	1,534,443	201,255	610,017	811,272	972,594	972,594	701,406
All Clerical and Kindred[c]	2,021,912	6,203,858	9,125,770	3,452,281	9,582,440	13,034,691	3,448,416	13,445,884	16,904,000

Sources: For 1960 and 1970, U.S. Census Bureau, 1970 *Census of the Population, Characteristics of the Population,* Vol. 1, Part 1, Table 221: Detailed Occupation of the Experienced Labor Force and Employed Persons by Sex: 1970 and 1960. For 1978, U.S. Department of Labor, Bureau of Labor Statistics, *Employment and Earnings During 1978: An Analysis,* Special Labor Force Report 218, Table 23: Employed persons by detailed occupation, sex and race.

[a] For 1960 and 1970, 14 years of age and over; for 1978 16 years and over.

[b] Rounded to nearest thousand.

[c] Including all clerical occupations listed in census, but excluding computer specialists.

by 1978 with the widespread use of dictating equipment, much of it semi-automatic and linked to telecommunications systems. Messengers increased more slowly than clerical occupations as a whole with the replacement of some paper communication with electronic transmission of information.

Yet, some jobs, which might appear to be rendered obsolete by automation, did not fall as much as might be expected. The work of file clerks might seem to be reduced by computerized record keeping, but apparently, the amount of paper grew sufficiently so that file clerks increased by 249% between 1960 and 1970, despite automation. By 1978, the number was reduced from the 1970 peak but was still 188% of the 1960 figures. It was thought that the spread of new technology such as word processing and semi-automatic dictating equipment, would eliminate the need for so many secretaries. However, the number of women employed as secretaries grew more rapidly than clerical work as a whole. One might speculate that the personal service and assistance rendered by secretaries remained critical to the operation of offices, and may have had particular importance in the expanding field of personal, professional, and business services. This latter factor may explain the disproportionate increase in the number of receptionists between 1960 and 1978; the Bureau of Labor Statistics projects a continued rapid growth in this occupation. Similarly, the number of typists continued to grow despite automated and semi-automated correspondence systems, due to continued expansion of communication.

Finally, one interesting indicator of change in the office structure was the disproportionate growth in the number of clerical supervisors. This pattern may be related to automation, in that with greater productivity and capital equipment invested in workers, there is greater concern with managerial control of the work process.

As these varying patterns indicate, the overall impacts of technology on the occupational structure are not easily pinpointed. Technology may improve productivity in certain areas, but the volume of activities can increase even more rapidly, thus offsetting declines in employment. Moreover, the mediating activities of clerical workers who link different operations and units within organizations through communications and record keeping may have increased as organizations have grown large and more differentiated into levels of hierarchy and separate functional units (Glenn & Feldberg, 1977). At the level of the occupational structure, with data from different industries—some of which are expanding and some which are shrinking, some undergoing change and others stagnating—it is difficult to separate out the impacts of technology from other effects.

A more precise picture is provided by looking at data from one industry. The Bureau of Labor Statistics conducted wage surveys of over 200 insurance companies in 13 metropolitan regions in 1961, 1966, 1971 and 1976. It is possible to examine changes in distributions in detailed occupational categories. Moreover, the actual activities of specific job categories are spelled out. The data on distributions are shown in Table 5.2.

One interesting overall trend, not shown in this table, is the increased differentiation of levels within occupations. For example, file clerks were differentiated into A and B categories based on degree of autonomy/responsibility in 1961. By 1966, there were three levels of file clerks. Similarly, in the electronic data processing (EDP) occupations, the levels of programming and systems analysts increased from two to three by 1971. These finer differentiations probably reflect greater rationalization of office procedures and more explicit definitions of duties.

As Table 5.2 shows, the fastest growing occupations were the "professional technical" EDP related ones—programmers and systems analysts, most of whom were men. Other computer related jobs expanded after 1961, but leveled off with improvements in the systems that increased capacities and speeded up processing. Computer operators leveled off in 1971, despite continued increase in the volume of automated processing. Keypunchers reached a peak in 1966; the replacement of cards with direct entry onto discs and tapes vastly improved productivity, while decentralized entry (via CRT at the point at which data were generated) reduced the reliance on data entry specialists.

The positions displaced included those involved in hand or mechanical processing: Tabulating machine operators had disappeared by 1976. Clerks involved in record keeping, transcribing premium payments, assembling records, calculating values on policies, handing remittances (assemblers, policy evaluation clerks, ledger clerks, and premium acceptors) were reduced, again despite increase in the volume of activity.

In the meantime, most other positions dealing with transactions remained at a stable level, including clerks in accounting and correspondence, while underwriters increased slightly in response to greater volume. Stenographers were rapidly disappearing and typists were becoming fewer; a new category of transcribing typist who worked with dictating machines made up for some, but not all of the loss of jobs. Although data on secretaries were not presented in 1961 and 1966, there was a substantial increase in secretaries between 1971 and 1976.

Examination of the gender composition of the various occupations and changes in that composition reveals a striking pattern. With the exception of keypunchers, all of the EDP related positions, including that of computer operators, were and remained overwhelmingly male. In contrast, most of the shrinking and stable positions that started out mostly female became increasingly feminized. Clerical positions that had a fairly sizable minority of men in 1961, had almost none by 1976. It has been said that jobs in which females are concentrated, or have recently been allowed to enter, are most likely to be displaced by technology (Hacker, 1979). There is no clear evi-

TABLE 5.2

Distributions of Employees in Selected Occupations in Offices of Life Insurance Carriers in 13 Selected Areas of the U.S.

	1961			*1966*			*1971*			*1976*		
Selected Occupants	***Men***	***Women***	***Total***	***Men***	***Women***	***Total***	***Men***	***Women***	***Total***	***Men***	***Women***	***Total***
Actuaries	983	83	1066	499	18	517	527	35	562	755	100	855
Assemblers	56	1294	1350	44	1022	1046	—	—	—	—	—	
Claims Approvers	—	—	—	391	338	729	402	491	893	363	866	1229
Clerks, Accounting	346	3174	3520	278	2702	2980	202	3530	3732	—	3887	3887
Clerks, Correspondence	782	1322	2104	427	1409	1836	228	1659	1887	—	1964	1964
Clerks, File	156	4637	4793	99	4494	4593	55	3725	3780	—	3629	3629
Clerks, Policy Evaluation	144	1397	1541	89	1045	1134	43	945	988	—	951	951
Clerks, Premium Ledger	25	1654	1679	16	780	796	3	553	556	—	263	263
Console Operators	131	10	141	417	53	470	—	—	—	—	—	—
Computer Operators	—	—	—	—	—	—	1765	481	2246	2104	257	2361
Keypunch Operators	2	3072	3074	5	3489	3494	7	3458	3465	—	2955	2955
Premium Acceptors	62	672	734	9	518	527	5	606	611	—	570	570
Programmers	503	112	615	1011	355	1366	1935	942	2877	2255	1242	3497
Secretaries	—	—	—	—	—	—	5	6413	6418	—	7532	7532
Stenographers	—	3586	3586	16	3867	3883	—	2336	2336	—	1286	1286
Systems Analysts	335	30	365	470	80	550	1465	345	1810	1914	790	2704
Tabulating Machine Operators	1789	1344	3133	992	895	1887	459	421	880	—	—	—
Tape Libraries	—	—	—	32	39	71	59	134	193	48	137	185
Transcribing Machine Typists	—	—	—	—	—	—	2	1021	1023	—	993	993
Typists	3	6389	6392	2	6514	6516	4	5327	5331	2	4098	4100
Underwriters	1156	404	1560	1154	410	1564	1119	555	1674	1004	657	1661

Sources: U.S. Department of Labor Statistics, *Industry Wage Survey,* 1961, 1966, 1971, and 1976.

dence in this case. The largest category eliminated by automation—tabulating machine operators—had slightly more men than women in 1961. As the occupation shrunk, the gender composition remained stable. Overall, however, it appears that with the exception of underwriting, the occupational structure in the insurance industry became more sex-segregated between 1961 and 1976, with the technical EDP jobs becoming a male province, while the routine clerical work became increasingly female.

In summary, taking data on both the broad occupational structure and the distribution in one industry, what can we conclude about the relationship between office automation and the occupational structure? First, overall, the continued expansion of clerical activity more than offset any savings in clerical labor; however, in insurance, a steadily growing industry that relies heavily on both automation and on clerical labor, there has been a reduction in traditional clerical jobs. The expansion of computer related occupations has increased the total number of jobs, but the workers displaced by automation do not appear to benefit, thereby. The new jobs are technical level and largely male. Women in routine clerical jobs are unlikely to be moving up into these jobs. Rather, a new layer of largely male workers has been inserted into the office, recruited from a different labor pool.

THE ORGANIZATIONAL LEVEL

Most studies at the organizational level deal with problems of transition and the effects of automation on worker attitudes.[1] They do not provide information on either the occupational structure or the organization of work in the organizations. Our review of this literature has revealed only two studies that present full before-and-after pictures of the occupational structure. We examine the information from these studies to see what light they shed on the degree of upgrading/downgrading and the extent of displacement/unemployment, including the differential impacts on men and women.

A case study of a large insurance company (Helfgott, 1966) shows that following the introduction of the computer, the number of clerks involved in tasks such as sorting, calculating, verifying, and maintaining records was markedly reduced. The reduction was so great that, even though many tasks that had previously been carried out in branch offices were now centralized

[1] The case study literature was published primarily between 1955 and 1968. Because the introduction of computers was expected to have dramatic effects on offices, many studies on the introduction of computers were conducted during this period. However, the chief concerns during this period were workers' reactions and the personal problems surrounding geographic and occupational shifts. The literature concentrates on these issues without providing basic information about the amount of change in occupation or the numbers of people shifted to different jobs, information that would provide a context in which to interpret the attitudinal data. As a result, we found this attitudinal data of little value for this investigation.

at the home office, there was still an overall decline in clerical employment in the home office.

In 1960, the company had employed 4,475 clerical workers, accounting for two-thirds of the total employment in the home office. By 1964, that number had dropped to 3,872 clerical workers—a loss of 603 positions. In addition, the 1964 figure included the new job of keypuncher, which represented 154 jobs (or 4%) of the clerical category. Thus, a total of 757 or 16.9% of traditional clerical positions had been eliminated. At the same time, the company had added new technical employees (programmers, computer operators, systems analysts) and increased the ranks of supervisory employees to expand that combined category by 275 jobs or 15%. The managerial staff also increased during this period, by 78 jobs or 20%.[2] By 1964 clerical workers had dropped to three-fifths of the work force in the home office.

How did these changes affect the sex composition of the office? The report does not provide data on sex distributions in different jobs. However, it implies that traditional clerical jobs were almost entirely female, as were the new keypunch jobs, whereas the growing technical, supervisory and managerial jobs were held largely by men.[3] Thus, between 1960 and 1964, there was a marked decline in female employment in the home office. By 1965, the author reports that women were down to 55% of the total work force, from over 65% 5 years earlier.

These changes confirm the patterns described for the occupational structure in the previous section; that is, while the dominance of women in the office labor force declined, the dominance of men in the upper levels of the office hierarchy increased. The technical staff was introduced into the office as a new elite, a separate stratum between clerical workers and managers that added another level to the office hierarchy. Thus, women clerical workers have become a relatively smaller group in a relatively lower position.

Helfgott's (1966) parallel study of a small, rapidly expanding insurance company found less dramatic changes, but the pattern was confirmed on all dimensions. In contrast, Rothberg's (1969) report on office automation in the Internal Revenue Service shows considerable difference. First, the total labor

---

[2] Increase in the size of the managerial staff following the introduction of computers has been noted by several researchers (Helfgott, 1966; Hoos, 1961). However, there is no agreed upon reason for it. Various reasons are suggested. Each worker produces more and, therefore, needs to be more carefully supervised; new kinds of jobs require new kinds of supervision and new layers of supervision need to be tied into the administrative hierarchy; and, as the remaining clerical tasks become "more qualitative" and "less quantitative," some of the work shifts to administrative personnel—this last according to one administrator of Inland Steel Company.

[3] It is interesting to note that most jobs directly related to computer processing, such as computer operators, are classified as technical, whereas keypunchers are classified as clerical. It may be that the presence of large numbers of women as keypunchers was instrumental in determining the classification, because all the predominantly female jobs in the office were clerical jobs.

force grew by 80%. Many traditional clerical occupations expanded. Examining and statistical jobs and sorting, classifying and filing jobs, which together accounted for 30% of employment under the old system, rose to 50% of employment in the new system. In addition, keypunch and other machine operators doubled in number. At the same time, the small category of secretarial workers declined and "posting" grew only half as fast as total employment. Furthermore, almost 30% of the labor force "chose" to retire or to quit during the transition period, and another 109 people accepted "downgradings" as an alternative to moving outside of the commuting area.

Rothberg (1969) provides no data on the sex distribution of these changes or the sex composition of job categories; however, preliminary data are available in a earlier report (BLS Bulletin #1364). In the IRS, as in the insurance companies, traditional clerical jobs were held largely by women (posting, checking, and maintaining records was over 95% female; examining and statistical was almost 90% female and sorting, routing, classifying, and filing was 80% female). Administrative positions were held largely by men (over 90% male). With the shift to the computerized system, total employment increased by 80%. Some traditionally female jobs accounted for a large percentage of the increase. However, the new, upper level jobs associated with electronic data processing were filled almost entirely (95%) by men. Thus, although women continued to predominate in the total work force, men's domination of the higher levels of the office increased.

Two other aspects of change in the occupational structure also are revealed by the studies of the Internal Revenue Service (see Rothberg, 1969). First, the changes created new barriers to upward mobility. Many of the high level, complex clerical jobs disappeared early in computerization. This is confirmed by studies in the insurance industry (Cummings, 1977; U.S. Dept. of Labor, 1965) and a study of the cost accounting department in a steel company (Weber, 1959). This change compressed the occupational hierarchy, cutting off occupational mobility for employees in certain departments or specialties. Similarly, middle level jobs that were previously viewed as training positions also were eliminated or changed in ways that reduced training opportunities. As a result, these jobs and the ones below them became virtual dead ends. Again, there is supporting evidence (Hoos, 1961; Sheperd, 1971). Second, with computerized processing, the Internal Revenue Service found that communication with clients became increasingly important. Of the clerical jobs that remained in the district offices, 49 were characterized as "public information and assistance," the same number as had been required under the old system. In addition, a new position called "data processing specialist" was created at the central data processing office "to more effectively handle taxpayer complaints and monitor the adjustment procedures and flow of forms . . . to ensure a correct and final settlement with taxpayers" (Rothberg, 1969, p. 29).

What light do these studies shed on the questions we asked? Despite the apparent differences in outcomes, we would argue that both suggest similar relationships between the use of computer technology and the occupational structure of clerical work. First, there is no evidence of upgrading of the clerical occupations as a whole. Although new higher level jobs are added to the office, these jobs are not clerical jobs and are not filled by former clerical workers. The new clerical jobs created in relation to this new technology are routine, relatively narrow jobs that do not lead to higher level jobs. Second, there is no evidence that the presence of higher level technical jobs in the office has an upgrading effect on either traditional or reorganized clerical jobs. If anything, there is some evidence that the opposite may be true: that as clerical work is reorganized, partly in response to the use of the new technology, it is narrowed, fragmented and limited, and some of the highest level jobs are eliminated. Third, in both cases the brunt of displacement/unemployment during the initial introduction of the new technology seemed to be concentrated in the jobs traditionally held by women. Other evidence suggests that this pattern reflects not only technological consequences, but also employers' beliefs that women will have higher rates of attrition (voluntary) and their consequent hiring of women to fill temporary jobs during the changeover (Hacker, 1979; Mann & Williams, 1960). The degree to which displacement/unemployment occurs depends on the kind of work involved, the personnel policies adopted by the organization, and the growth of the organization's activities, and thus, is not readily predictable.

Although the average rank of an office labor force (a structural property of an organization) may rise following the introduction of electronic data processing it cannot be argued on the basis of these data that the clerical work force is upgraded.

Each job category must be analyzed separately. The limited attention to the specific details of the consequences of office automation, given the relatively large number of studies covering the subject, suggests that researchers systematically ignored this issue—perhaps, because of the long-standing bias that women workers, particularly those in offices, were not real workers whose work situation deserved/required careful thoughtful analysis. The information felt to be most necessary was an assessment of their attitudes toward change and the factors that increased their willingness to cooperate with management in running automated offices.

THE WORK PROCESS

Finally, we approach the issue of the managerial control or worker autonomy at the work process level. It has been argued (Blauner, 1964; Sheperd, 1971) that automation makes it possible to bring together into a single job functions that were separated out and divided among several jobs. The presumption is that the job that results from the bringing together of these activities is a more integrated whole; one that gives the worker greater autonomy and reduces alienation. Unfortunately, this debate has never been properly assessed in relation to clerical work because the major published investigation (Sheperd, 1971) deals with the topic solely by examining the workers' attitudes. If workers appeared less alienated on a number of dimensions, Sheperd assumes that their jobs in automated offices are more integrated and that they are autonomous, without looking at the actual content or organization of those jobs. Therefore, we turn here to our own case materials to compare two apparently opposite work situations. In the first, we describe a job organized to carry out subtasks of a computer-assisted operation. In the second, we describe a job made up of once separate jobs that can now, with the information available via computer, be done by one person.

This description of claims processing in a large insurance company comes from field observations and discussions with managers and workers in these units. A few details have been changed to preserve anonymity.

Because claims are the central business of an insurance company, the processing is surrounded by "control" measures. Although these are not fully described here, we want to point out that each time a batch of claims is moved from one unit to another, it passes through the hands of a control clerk who notes in a log book where the batch has come from and where it is going next. Once the claims are grouped, they are sent to data entry operators who call up the claimant's permanent file from the computer on a screen. The clerks check the account to be certain the claimant is "paid up" and, therefore, eligible to enter a claim. Then they fill out a form using code numbers to describe the kind and level of coverage the claimant has. After going through the control clerk, the claim is reviewed by the claims examiner. She uses the coverage codes to look up the details of the claimant's policy in a policy manual, checking to be certain that the kind of claim being made is covered by the policy in force. Then she checks the expenses claimed against a manual of standard rates for different kinds of expenses. Because the claimant's description of the reason for a claim and the category of expenses rarely coincides with company examples, the claims examiner must be familiar with the policies and know what kinds of variations are permitted or possible for different kinds of policies and different kinds of services. She examines the policy number for evidence of any riders indicating special conditions. Then the claims examiner totals the expenses claimed as a check on the claimant's calculation. Finally, she fills out a code form that summarizes all the information and adds internal company codes required for processing. Then the claim passes through the control clerk, and on to the data entry clerk who now enters the information from the coded forms into the computer so that a check will be issued to the claimant. The claims examiner adds a code for her name to each form completed. In that way, her productivity is tallied as the claims are entered into the computer and errors are evident in the claims that are rejected by the system.

The job of the claims examiner is one part of a partially automated operation. Although the examiner needs a certain amount of knowledge about company policy and records, she cannot complete the information required for processing the claim on her own. Vital parts of the information are supplied by the data entry operator and important checks are performed by the control clerk during the logging in and out operation.

In contrast, the clerk who handles customer complaints, inquiries, and requests for service in a public utility does all phases of the work. Before the computer system was fully developed, customers called separate departments for specific types of questions. Each clerk provided information on a specific area: sales, bills, service/repair, etc. Separate files were maintained in each office and a complex question might require several phone calls. With computerization, the files were centralized. Now, one set of clerks, using computer terminals, has access to all the information. The clerks who answer these phones can look up a customer's account and explain a billing problem, can set up appointments for service calls, or can tell a customer the balance due on a special service. In addition, the clerk can enter into the terminal any new information provided by the customer: the time someone will be home for the service call, the type of problem, a change in address or other billing information. The job is more "integrated" in that the clerk does more tasks than before. Management is looking forward to extending the clerks' work further. Clerks are being asked to enter routine billing information through their terminals during slack periods of the day. In addition, several monitoring systems keep track of the work done by each clerk. Machines monitor the number of calls answered and missed on each line, while the clerk's code, entered into the computer with every transaction, provides a tally and an error count on information processed. The clerk has more task variety than previously, but operates in a more rationalized system in which each part of the job must be done exactly according to specific patterns and within a specified time frame. Furthermore, although she has access to detailed information, the clerk's knowledge of the company procedures in each area is less complete than previously. Finally, the clerk is now seated at her desk/terminal throughout the day.

These are two apparently different types of jobs. One is a fragment of claims processing. The other is the complete job of customer representative. Yet, in both cases, there is a high degree of rationalization and very little evidence of autonomy. If anything, the use of the terminal by the "integrated" worker, the very instrument that makes it possible for her to do a "complete" job, also makes it possible for management to structure the employee's work more completely and to keep closer watch on her activities in an unobtrusive way.

CONCLUSIONS

The difficulty of drawing conclusions from the historical record on office automation stems from two kinds of bias. One is sexist bias: the general cultural bias that leads researchers to ignore or dismiss as unimportant problems faced by employed women in their jobs (Feldberg & Glenn, 1979). The other is technological/managerial bias: an interest group bias that, in this particular historical period, led researchers to view computer technology as "progressive" and promising and workers as obstacles to the fulfillment of that promise (Noble, 1977). As a result of these biases, researchers have never been able to resolve the debates or move beyond them and have continued to jump from level to level in seeking explanations.

Sexist bias can be seen in the treatment of both upgrading and displacement. Because women were not expected to move up the occupational ladder, researchers looked most closely at the implications of new technology for men's jobs. Where they found men experiencing upgrading, they interpreted the finding as representing the experience of all the workers in the office. Appropriate average statistics were developed to measure this upgrading, whereas the remaining jobs in the office, those held largely by women, were ignored. Similarly, the attempt to show that displacement was not inevitably a result of automation, led researchers to miss the extent of displacement experienced by women. Indeed, the bias was so strong that Helfgott (1966) did not calculate the total number of traditionally female clerical jobs lost and Rothberg (1969) failed to see that early retirement and quitting might be interpreted as hidden forms of displacement. In this context, neither individual researchers, nor government agencies saw the importance of these issues for women clerical workers.

The technological/managerial bias in this work is equally prevalent. It reinforces sexist interpretation of upgrading and displacement, but adds an additional note, as in the treatment of "worker autonomy." The promise of computer technology was that it would make life and work better for everyone through increased productivity coupled with the elimination of "bad" jobs. Technological experts and managers saw the major problems as coming from employees' resistance to the new technology. Because most researchers shared the belief that computer technology was "progressive," they, too, were interested in workers' attitudes toward the new system. Thus, they concentrated their studies on the questions that reflected the concerns of managers, and interpreted their data from a managerial perspective. The most interesting evidence of this managerial bias is the lapse of such studies following widespread acceptance of computers in offices. Once the issues of prime concern for managing the system receded, researchers turned their attention in other directions. As a result, we have little basic information on the changing conditions of work in computerized offices from the first introduction to the present.

Given these limitations, is there anything we can say about the transformation of clerical work in the post World War II period? As we see it, the transformation in this period continues many of the trends seen in earlier periods. The clerical occupations, as a whole, have continued to grow (although the rate of growth has slowed) and, simultaneously, in the process, clerical jobs have undergone further rationalization. Although designed to take advantage of the particular capabilities of the computer, this rationalization has not been fundamentally different from that of the 1900–1930 period. In both cases, rationalization increased the detailed division of labor, subjected workers to closer scrutiny and transferred decision making away from clerks (Glenn & Feldberg, 1977). In the earlier period, policy making and control of the work process were transferred from upper level clerical workers to an expanding stratum of middle level managers. The recent reorganization of the work process around automated systems is removing the procedural decisions from female clerks and placing them in the hands of a new, predominately male technical staff who design the work process. The continuation of these trends in a period of major technological change suggests that, although the technology per se is an enabling and shaping factor, it is its use within existing social patterns, such as gender hierarchy, which explains the direction of its impacts on the transformation of clerical work.

ADDENDUM

In the 5 years since this chapter was written, the literature on office automation and the reorganization of clerical work has grown enormously. But the debate on the implications of these changes for women clerical workers is still unresolved. We cannot yet say with certainty how many jobs will be changed or lost and what kinds of clerical jobs there are likely to be.

In part, the lack of resolution reflects the dynamic state of changes in the economy and in the use and availability of new technologies. Changes in the economy and the continued growth of service industries have meant continued increases in the number of clerical jobs. At the same time, the spread of new technologies has been somewhat slower than predicted. Together, these trends and others have meant more clerical jobs at a variety of levels than we might have expected.

A notable example is the continued increase in secretarial jobs, contrary to all predictions. Trying to explain this increase points to several outstanding questions. Is the category meaningful? Do jobs classified as secretarial have enough in common to be grouped together? In what ways are they similar to or different from an older concept of "secretary," which was the job predicted to decline? Perhaps the increase in secretarial jobs masks the emergence of a new type of clerical work, as yet undefined. Perhaps the symbolic functions of secretaries loom larger now that word processing permits sending hundreds of individually typed letters. Or perhaps those who predicted the disappearance of the secretary did not understand her "invisible work" (Murphree, 1984)—the real but unacknowledged work that secretaries do and facilitate. One boss said that he works more efficiently when his secretary is at her desk because he knows someone is waiting for the work and is ready to complete it. The secretary may be a prod, almost a supervisor, as well as an assistant to the boss. Trying to explain this unexpected increase takes us behind the numbers to understand the complexities associated with the reorganization of the office.

In part, the lack of resolution reflects the continued confusion of what this chapter identified as levels of analysis. What is true for the nation as a whole may not be true in particular industries or particular firms. Likewise, what is true of the office as a whole, may not be true of particular groups of office workers. Together, these statements suggest that to understand the meaning of the changes in clerical employment requires a more detailed picture of differential implications. We must unfold the category "clerical workers" to see what is happening to, for example, young, high-school educated black women in the insurance industry or older, high-school educated white women in state and local government.

More recent analyses have looked in detail at the intersections of sex, race, age, level of education and industry (Applebaum, 1984; Baran, 1984; Baran & Teegarden, 1984; Noyelle, 1984). They have found, for example, that as entry level clerical jobs (such as keypunching) in the insurance industry decline, young, less-educated women, especially black women, face new barriers in entering clerical occupations. At the same time, the growth of higher level jobs (such as the telephone representative) means larger numbers of more interesting, although not necessarily well-paid, jobs for college-educated women. As the jobs above and below both types of positions disappear, clerical workers appear to have even more limited prospects for mobility than we expected. Thus common consequences may exist side-by-side with differential effects. Only by looking at several levels of analysis can we understand both outcomes and their relation to each other.

Even more important than the lack of resolution is the shift in terms of debate. Researchers are no longer trying to characterize the changing technologies as "good" or "bad," as "upgrading" or "deskilling" the entire occupation. They have recognized the differential implications of any set of changes. It is no longer controversial to argue that some jobs that are being created are "lower level," more boring, or at least no better than jobs that were eliminated, whereas others are "higher level" and more interesting. The most recent analyses of the insurance industry (Applebaum, 1984; Baran, 1984) suggest that the lower level, already deskilled jobs are now being eliminated (built into the programs) in the most sophisticated reorganizations of work, whereas new, higher level clerical positions are being created. Not surprisingly, much attention has focused on the new better jobs. One hypothesis, recently advanced, is that these more skilled, more integrated jobs, resembling the one described at the end of this chapter, will constitute an important, growing segment of the clerical occupations.

Here the debate must be sharpened before it can be resolved. We need to know how many of these "better" jobs will be created and how large a part of clerical employment will continue to be found among narrow, more routine jobs. We need to know in greater detail what both sets of jobs will be like and how they will fit into the organization of work. For example, the routine jobs may continue to be full-time permanent positions, or some portion of them could be added to the present one and one-third to two million temporary clerical jobs, as happened earlier with keypunching. Because such jobs typically have no job security, fewer benefits, and even more limited opportunities, such a shift represents a decline in working conditions even if the work remains the same.

For the new, better positions, we have limited information about what the jobs will involve and how they will fit into the organization of work. In some industries, it appears that these new jobs will primarily serve customers; in others, they appear to combine the traditional clerical functions with some of the "professional" activities formerly belonging to other groups of workers. In still other places, the quick rise of the personal computer means that some clerical tasks are being performed by "professional" employees in the course of doing their jobs, which may remove some clerical functions from clerical jobs.

Furthermore we do not know how these different possibilities are developing—which are becoming widespread in which industries. Nor do we know how clerical workers will gain access to and fare in these jobs. Recent research claims that these jobs will require more skills—at least more sophisticated reasoning ability, a wide understanding of company policy and procedures, as well as some technical skill in dealing with computer programs. How will these skills be gained? Will educational credentials be the proxy for them, restricting access to those with associates or bachelors degrees? Will the more specific skills acquired in doing this work be recognized, so that these jobs are both better paid and the basis for upward mobility or, as some analysts have predicted, will these jobs become new enclaves, new dead-ends for middle-class women? Finally, how much autonomy will workers have in carrying out these jobs? A "higher level" job can require more skills, even pay a higher salary, yet still be limited in autonomy, closely monitored, afford little discretion, and be tied to production quotas. We need to know more about such jobs in order to decide in what ways they are or are not better than the jobs they replaced.

The debate about which workers are getting which jobs is changing in a far more dramatic way. As previously discussed, the study of differential effects has progressed in this area too. Researchers now identify groups by the intersection of race, age, and education with gender, pointing out effects for specific groups e.g., black high-school educated women. Looking at these specific groups in relation to the emerging job categories leads to a disturbing new question: Which women will find jobs in the clerical occupations? National statistics show an increase in the proportion of black women in the better clerical jobs, as in the occupation as a whole, between 1970 and 1980. Yet factors such as the decline in entry level positions in insurance (such as key punching/data entry), new educational requirements and the movement of jobs away from the inner cities to suburban centers and small towns—and even overseas or "offshore"—could reverse this trend.

GENERAL CONCLUSIONS

The relationship between the changing organization of offices and clerical work is still in flux. From research over the past 5 years, some elements of this relationship are clearer, whereas others are revealed as more complex. For example, it is now clearer that technological change does not itself determine how offices will be organized and what clerical jobs will exist. Yet the relationship among changes in the economy, the use of new technologies and employment is more complex than anticipated. It varies not only by industry but within industry, reflecting differences in both the nature of particular businesses and the historical context in which the managerial practices were established. The result is both common and contradictory outcomes at different levels and for different groups of workers.

Trying to unravel those complexities reveals three sets of outstanding questions. First are those that ask what kinds of clerical jobs are emerging. These cover the nature of the jobs, the conditions of work, the quality of work experience, and its place in the employing organizations. Second are those that ask about numbers. Here the issue is how many, of those kinds of jobs, will there be. Third are those which ask about the relationship between the jobs and the workers. These address the issues highlighted by the concept of differential effects: which workers are being recruited for and excluded from which jobs, according to gender, race, age, education, and related social characteristics.

Because such critical questions remain outstanding, it is especially important for a wide range of constituencies to take part in uncovering new information and influencing public policy.

ACKNOWLEDGMENT

This article is a revised version of "Technology and work degradation: Effects of office automation on women clerical workers" that appeared as chapter 4 in Rothschild, J. (Ed.), *Machina Ex Dea: Feminist perspectives on technology.* New York and Oxford: Pergamon Press, 1983 © Joan Rothschild 1983.

6 Computers and the Automation of Work

Philip Kraft
Department of Sociology, State University of New York at Binghamton and Center for Survey Research University of Massachusetts-Boston

ABSTRACT

Although programmers often are thought of as the prototypical knowledge worker in the post-industrial society, this chapter argues that programmers and other workers in the software field are caught up in traditional labor processes and social relations. Computers have replicated the relations of the shop floor, not transformed them. It is not the technological content of the work, but the differences between manager and managed, between conception and execution, and between men and women that characterize software work. The people who do the programming are not the ones who decide what to program or what machines to buy or languages to program in. Software work has become less skilled as programmers are subject to efforts to simplify and standardize their work. Although programmers often are thought of as professionals who are rewarded for their technical expertise, managers of programmers, who are generalists, are paid far more. And software as an occupation has replicated traditional occupational status and reward hierarchies, in which women do more routine and more repetitive work and are paid less than men regardless of the work that they do.

The computerization of work is usually described as a revolution, not only in the way we make things, but in the structure of the larger society. I take another position: Although computer technology is dramatic, its application is not. The computer is a revolutionary tool that will have largely conventional effects in both the workplace and in the society as a whole. In particular, the computer will help reproduce, not transform, traditional relations between manager and managed, between those who design and plan and those who carry out the designs and plans of others, between those who have jobs in which experience and judgment are valued and those with jobs in which experience and judgment are irrelevant. Computers will reproduce the divisions between people who are paid well and have reasonable job security and people who are paid badly and have little security. In short, instead of replacing bad jobs with good jobs, computers are likely to create some good jobs, some bad ones, and eliminate a lot of others.

This is not an abstract issue. Much has been made of the gloomy future awaiting a society that relies on old-fashioned smokestack industries. Jobs, prosperity, and even political stability are said to be at stake if we don't make the transition to something that has been called an information society, service society, or, to use an older and quainter phrase, a post-industrial society. Our future, according to the post-industrial scenario, requires phasing out traditional blue-collar work—and workers—and replacing them with high technology work and "knowledge workers."

THE COMPUTERIZATION OF COMPUTER WORK

In an information society the ultimate knowledge workers are computer software specialists. Programmers, as they are usually called, are at the leading edge of the computer revolution. With these prototypical knowledge workers we realize our most extravagant claims or our worst fears about computerization. What skills do programmers have? How well are they paid? What are their career prospects? How much control do they have over their jobs and their work? How different is their work from older and more conventional work?

My colleague Steven Dubnoff and I (cf. Kraft & Dubnoff, forthcoming) are completing a 3-year study of software specialists. Our findings show little support for the post-industrial thesis, nor do they suggest a dramatic role for the so-called knowledge workers. We found instead a continuation of a very old process—the mechanization and automation of work—using a new tool, the computer. We also found that computer-based automation has permitted the elimination of work that up to now had been exempt from older forms of automation. These are the same "knowledge" jobs on which the post-industrial theorists have placed so many hopes.[1]

[1] There is surprisingly little information about programmers. Bureau of Labor Statistics and Census data offer a skeletal picture at best. Industry and academic studies of software specialists invariably have been based on convenience samples of limited generalizability. By contrast, ours is a probability sample of the occupation as a whole. It is based on interviews of 667 software specialists of all kinds: programmers, systems analysts, and software engineers, as well as managers, entrepreneurs, and "contractors." All respondents were interviewed using a highly structured field interview. A subsample of the original respondents was interviewed a second time using an unstructured and open-ended schedule. We believe our sample is broadly representative of software workers rather than of a particular firm, industry, or specialization. The sample and survey techniques are more fully discussed in Kraft and Dubnoff (1986).

Specifically:

1. Programmers have been subject to efforts to simplify and standardize their work and make them easier for managers to control.

2. Software work is polarized. Some software specialists do creative, challenging work, whereas others have routine, repetitive, and boring jobs. The routine and repetitive jobs are vulnerable to computerization.

3. As a result of (1) and (2), software work, on average, is becoming less, not more, skilled.

4. As computers are being used at a faster rate, the rate of increase in the number of computer software workers has declined. In some parts of the country and in some specializations an accelerating supply of programmers has caught up with a slowing demand, i.e., some programmers are unemployed.

If we consider that computer software workers are not even 2% of the United States workforce, the demand for these mind workers seem to have peaked before it ever got off the ground. If we expect to replace traditional smokestack workers with new knowledge workers, these are not good signs.

We have other reasons to be less optimistic than the information society theorists:

1. In 1981, mean annual earnings of software specialists exceeded $31,000. The figure is misleading, however, because managers as a group had substantially higher earnings than the rest of the sample. When managers and supervisors are excluded, the mean earnings are a more modest $26,300. This is about the same as garbage collectors earn in large cities.

2. In a field that has tried to establish itself as a profession, most specialists are required to formally account for their time on the job, usually in units of an hour or less. Some are required to sign in and out. Some even use time clocks. Women are much more closely managed in this respect than men.

3. The key to financial success in software is not to specialize and not to get too technical. People with the highest paying software jobs spent some but not most of their time on technical matters, e.g., designing specifications for software systems. Similarly, they spent some but not most of their time directly supervising others. Instead, they spent most of their time making general decisions involving hardware and software purchases, planning, and on other managerial tasks.

The highest paid specialists are not specialists at all, but generalists. In spite of the talk about the financial rewards available to highly skilled technical specialists, in software work as in the rest of the corporate world the rewards go to managers, not to technicians, engineers, or scientists.

WOMEN IN SOFTWARE WORK

We have still other—and I think more compelling—evidence that computer-based work does not automatically alter the social relations of the workplace. That evidence comes from the role of women in software production.

Women occupy a special position in software work. Compared to other technical fields, such as electrical engineering, they are present in large numbers. The high percentage of women in software jobs often is cited as evidence that software work is not merely new, but different.

There are good practical and theoretical reasons why women should find software both rewarding and relatively free of gender discrimination:

- *Computer-related occupations are new.* The modern computer is barely 40 years old. The industries that developed from its application, especially the design and writing of computer programs, did not enter the world with a tradition of discrimination against women. Similarly, there are few if any restrictive practices that exclude women from workers' organizations, such as unions, if only because there are no software unions.
- *There has been a steady sellers' market for the services of skilled software specialists.* Even if potential employers are reluctant to hire women, the shortage of trained and experienced programmers has been a powerful inducement to overcome such attitudes.
- *Discrimination is against the law.* The growth in the software occupations coincided with the implementation of affirmative action and equal employment opportunity regulations.

There are equally compelling reasons to expect software work to reflect the traditional relationships between men and women in the workplace. Whether we like it or not, women have played a particular role in the United States workforce. Usually women are in less desirable jobs than men or get paid less than men for the same work. Women often are hired only when there is a shortage of men and are then fired when men are available. Women managers usually manage other women, not men. This follows an old and hard-to-change practice that workers are managed by people of the same or higher, but not lower status.

These and other factors suggest that there are really two different workforces in the United States, one made up primarily of white males, the other made up of everyone else.

The role of women in software work can serve, therefore, as a social litmus test of the claims made for the impending information society and the knowledge workers who will populate it. If women find software occupations both rewarding and accessible and if they are treated about the same as their male colleagues with comparable credentials, this would support the claims of the post-industrial theorists. In particular, it would indicate that social change, including change in the workplace, is technologically driven.

If, on the other hand, women occupy the same relative position in software work as they do in other occupations, then we are obliged to treat computer-based occupations (and perhaps other high-tech occupations as well) like older and more conventional jobs. This means that there is not only a division of labor but a stratification of jobs, complete with hierarchies of power and rewards. Some jobs will be interesting and challenging and well paid, whereas others will be routine and monotonous and pay relatively poorly. In short, we would have to conclude that social relations in the workplace flow from social imperatives, not from technological innovation per se.

Now, are women treated any differently in programming than they are in more traditional occupations? Have women software specialists escaped the "59 cents syndrome," in which they are paid about 59¢ for every dollar a man earns for the same work? Are all fields and industries as open to them as to men?

The short answer is no. Our findings indicate that software as an occupation has replicated traditional divisions and systems of rewards. Instead of revolutionizing the traditional social relations of production, software has reproduced them.

Let's begin with earnings. Women software workers are paid considerably less than men. The average earnings for women software specialists were $27,666 in 1981, in contrast to $36,510 for men. Part of the reason is that women are concentrated in the worst-paying industries that employ programmers, such as insurance, finance, real estate, and communications.

At first, this seems to be another of those interesting chicken-or-egg question: Are women concentrated in the lowest-paying industries or are those industries low-paying because they employ a lot of women?

In fact, cause and effect are easy enough to sort out. Even when women are in high-paying industries, they are paid less—usually substantially less—than their male counterparts (see Fig. 6.1. and 6.2).

When we control for every reasonable variable, on average women can expect 85% of what men with comparable credentials and experience earn for the same work. The unexplained difference is probably gender discrimination.

It is tempting to say that women have done so poorly, in comparison to men, because they have not been in the field very long and only now are entering in substantial numbers. Neither is true. Women have been involved

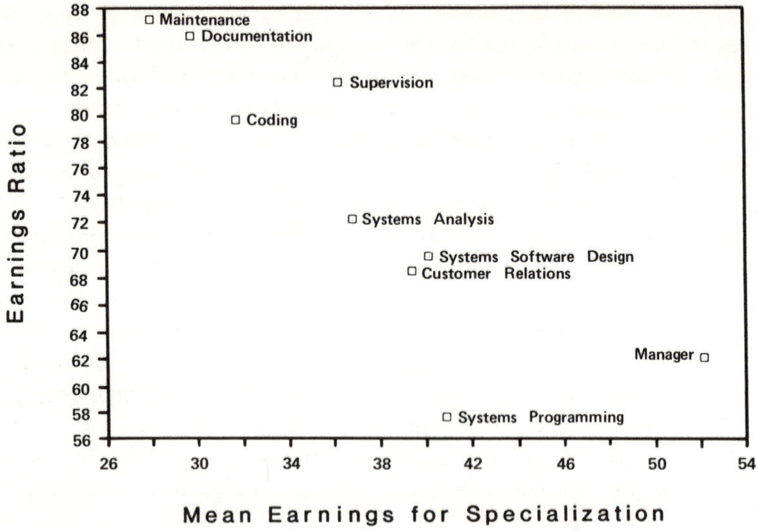

FIGURE 6.1. Ratio of female to male earnings.

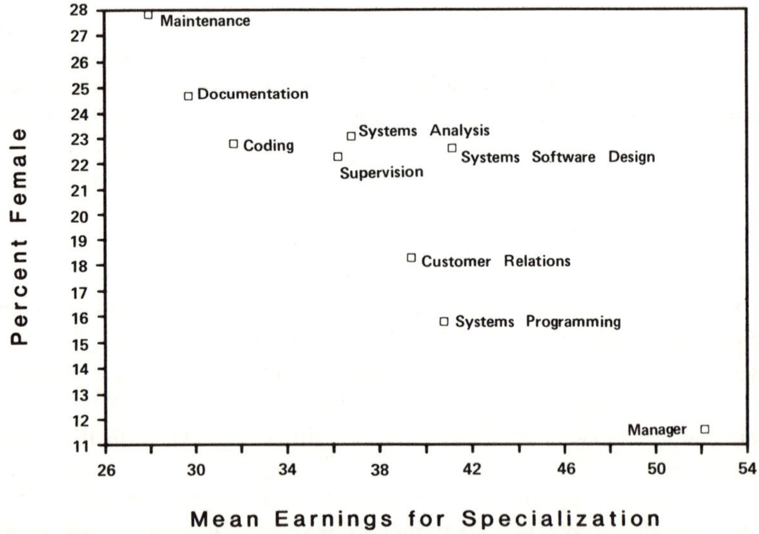

FIGURE 6.2. Percent female of specialization.

with computers from the beginning. Everyone knows that Ada, Lady Lovelace, helped program Babbage's analytical and difference engines at the beginning of the 19th century. Fewer know that women programmed America's first electronic computer, the ENIAC, during World War II.[2] The fact is that the first American programmers—nearly all of them—were women.

More to the point, the percentage of women in software jobs has remained essentially stable for at least 20 years. Although the percentage has now begun to edge up from about 20% to about 25%, women always have been a significant minority in the software workforce. This suggests to me that women have had enough time to move up the career ladder, but for some reason haven't.

Part of the explanation can be found in the career opportunities women have, particularly in management. Women are COBOL programmers, maintenance programmers, technical writers, and, for lack of better terms, the wheel greasers and go-betweens mediating between engineers, salespeople, clients, and users. These are, collectively, the bottom of the software pay ladder. Although we have not finished this part of our analysis, our preliminary findings suggest that these specializations are also dead-end jobs with respect to significant career opportunities, especially in management jobs.

Even when women have management jobs—and nearly 22% of them do—they are not the same management jobs men have. Nor is the pay the same.

Women are more likely to be straw bosses than high-level managers, supervisors rather than executives. I have already pointed out that women are likely to manage other women rather than men, a sure sign of low status.

The pay situation for women managers is astounding; there is no other word for it. Remember that managers earn substantially more than other software workers. It is in management jobs that the relative pay discrepancies between men and women software specialists are the greatest. The higher women go up the management ladder, the greater the gap in male–female earnings. To put this in the most dramatic terms, on average the lowest paid male supervisor makes more than the highest paid female executive. For women, the worst situation is in small firms, those employing fewer than 100 people, where women upper level software mangers earn half as much as their male counterparts. In large firms (>500), women managers earn as much as 70% of their male colleagues (see Fig. 6.3).

MIDDLE MANAGERS

When work is slated for standardization and eventual automation, it is sometimes given to middle managers first. Actually, such work is often given to

[2] One of ENIAC's first uses was to perform the complex calculations required by the Manhattan Project. Thus, the "ENIAC girls," as they were inevitably called, were responsible for ushering in not only the computer age, but the atomic age as well. For all her fame, Lady Lovelace could not claim as much.

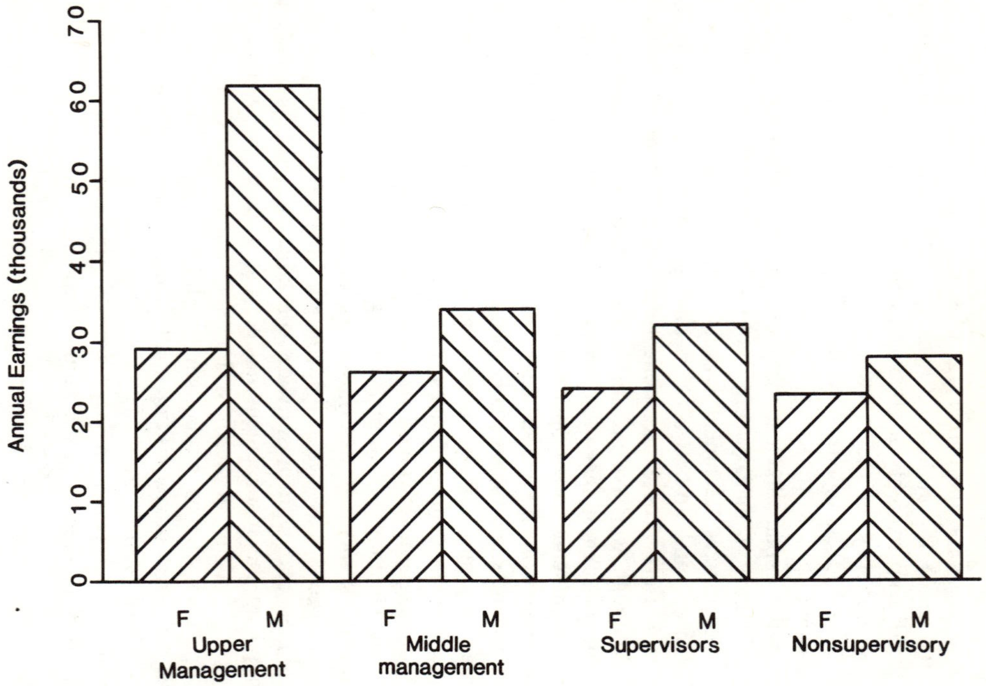

FIGURE 6.3. Annual earnings by position and gender.

managers who have been assigned new machines: hardware machines like personal computers and software machines such as Lotus 1-2-3. There should be no mystery about this, only some concern for the job security of those middle managers. Tools and machines such as Supercalc or Multiplan are devices that allow middle-level managers to do what they used to do with the help of clerks, accountants, and technicians. The point of a Visicalc or a Multiplan is to collapse and automate some or all the work performed by those various employees. In the initial stages, it is reasonable for middle-level managers to do the job; this, after all, is what they are supposed to manage and these are supposed to be the areas of their experience and expertise. After a while, however, we can anticipate that the hardware and software machines will get "smarter" while the people who run them will need less experience and less expertise. At this point, the middle manager and his machine will be replaced by a clerk and her machine.

This process is better understood if we recall a similar process that took place in health care. When mercury thermometers were first introduced they were kept in velvet-lined cases in the doctor's office safe, along with the narcotics. No one was allowed to touch the new tool except the physician. Today, there is hardly a physician who will admit to recognizing a thermometer, nevermind actually taking a patient's temperature. Even nurses have disdained the tool and have bequeathed it to nurses' aides. The process has been repeated with x-rays, cardiograms, echocardiograms, and even CAT-scanning machines. The one constant has been the division of labor, fragmentation of work, and hierarchy of power, control, and rewards in the delivery of health care.

All of this is consistent with the most fundamental imperative of modern capitalism, namely the substitution of unskilled labor for skilled labor and the substitution of capital for both. Managers are not only skilled labor, they are the most expensive skilled labor. There is no theoretical or practical reason why they should be exempt from the trends I have just described.

Casparis (1983) has pointed out that the history of industrialization, of which automation is only the latest stage, is also the progressive substitution of formal, engineer-designed quality control for the experience, discretion, and judgment of the individual workers. Some managers and management theorists understand this process quite clearly. They also have begun to take the next logical step: the automation and elimination of the quality control functions of middle-level managers and engineers.

Some readers will have read of the emergence of the so-called hourglass shaped organization, one characterized by a few executives and research and development specialists on the top and many clerks and operatives on the bottom. The middle, which used to be filled with managers, has been replaced by a communications channel. This view of organizations—perhaps we can call it the ultimate fantasy of upper level managers—is most clearly and matter-of-factly expressed by Franco de Benedetti, the co-managing director of Olivetti (1979):

> Standardization is the necessary prerequisite to the subsequent mechanization and automation of the productive process. . . . Data processing is . . . a continuation of a story which began with the industrial revolution . . . it is basically a technology of coordination and control over a labor force, the white collar workers, which [Scientific Management] does not cover. . . . In this sense, edp [electronic data processing] is in fact an organizational technology and, like the organization of labor, has a dual function as a productive force and a control tool for capital.

We have witnessed, according to de Benedetti, "a shift from hierarchical structure towards a polarization, with the elimination of intermediate groups and . . . a centralization of information and decisions prejudical to middle-managers."

COMPUTERIZATION AND SKILL

Finally, what about job content and skills?[3] This is a complex issue for two reasons. First, software work is still relatively new and continues to evolve. Second, skill in general is a complex concept and has not yet been adequately analyzed. In particular, the social component of skill is a fundamental issue that has been largely ignored.

Having said this, we can point to those results of our survey that deal with skill. One of the characteristics of "knowledge" work is that it is by definition intellectual labor. The core of intellectual labor is judgment and discretion. But I also have said that the overriding necessity of our production system is to eliminate human discretion and to transfer control to machines. In software work, the anomaly has been confronted in a straightforward and, I would say, an entirely predictable way: This prototypical mind work has been broken down into head and hand work and then the

---

[3] Harry Braverman's (1974) landmark *Labor and Monopoly Capital* renewed the long-dormant discussion of the nature and distribution of skill. Braverman warned of the "degradation" of skill, but tended, ironically, to accept management definitions of skill as a measurable individual quantity—almost a commodity—which managers and workers struggle to possess. This perspective has been criticized by researchers who otherwise share Braverman's view of worker/management relations. In particular, Kusterer (1979) has correctly pointed out that skills, like tools and commodities, are social and thus collective products as well as individual traits. As collective attributes, they embody the give-and-take, co-operation, and conflict of the workplace. In a workplace characterized by the struggle for control—which I think it is fair to say, characterizes most American workplaces—the definition of skill is at least in part the prerogative of the people in charge. This is not a new insight. Guilds understood this, unions used to understand this, and today managers and engineers understand this best of all.

components parcelled out to different workers. The trick has been accomplished in a way that will come as no surprise to management specialists. Some people make general decisions about design, hiring, expenditures, schedules, and so on, while other people do the work. Programming, in short, has been industrialized (see Kraft, 1977, chapter 3; Kraft & Dubnoff, in press).

There is an office where policy is made and selling is conducted. A second tier of managers supervises and evaluates designers who in turn specify and oversee the labor of the technical "detail men" who actually make the systems and applications software. Finally, there is a group whose function it is to grease the wheels and keep the customers happy. The people who make the decisions and set the specifications hardly ever get their hands dirty writing code or manuals or testing software. Similarly, the people who actually do the software work almost never make decisions about the hardware or software they use, the programs they will work on, or the direction of their organization. This is, I think, as clear a picture of what Braverman (1974) called the separation of conception and execution as one is likely to see, except that it has taken place in an intellectual occupation. Conception and execution have been separated in mind work.

CONCLUSION

Our findings indicate that in this most high tech of high-tech occupations the traditional differences between manager and managed are the most salient characteristics of software work, not the technological component of the work, not the emphasis on mind work, not the high capital–labor ratio, but an entirely conventional division of labor and separation of control and execution.

Computers have replicated the traditional relations of the shop floor, not transformed them. On the contrary, computer work itself has been transformed and now looks remarkably like conventional work. The shift, appropriately enough, has been reflected in techno-speak, which has transformed "programming" into "software production" (and, it should be noted, also transformed "idiot proof" into the "user-friendly professional workstation"). What is remarkable about the software shop is not how different it is from the machine shop, but how similar.

Ironically, the machine shop and other traditional industries may be more advanced in at least one respect: There are sizeable numbers of blacks and other minority Americans in the auto and steel industries. The same cannot be said of computer software occupations, which are 95% white.[4] The

[4] The breakdown in our sample is white: 95%; black: 1%; other: 4%. This is essentially the same as contemporary (1983) BLS findings.

difference is instructive and relevant to the issue of whether social change is technologically driven or vice-versa. The job opportunities for minority Americans in the steel, auto, other traditional secondary industries—and in garbage collection, for that matter—did not come about because of the introduction of mass production techniques. Women and people of color found jobs—relatively well-paid jobs with a degree of economic security—because they agitated for those jobs and demanded a basic level of pay and security. Struggles in the workplace, in the union hall, and in the larger political arena gradually opened more jobs and more industries to people who had been excluded from them because of prejudice or greed or both.

Although there is a technological component to these changes—obviously the opportunity to work in an auto factory requires the existence of a factory that makes automobiles—the moving force in this process has been social and political.

For these reasons I think the lessons taught by our much-maligned "mature" industries are still worth learning. Chief among these lessons are

- Technological innovation is not introduced into a production process because it will upgrade skills or make work less onerous or make jobs more interesting. Technological innovation almost always takes place because under our system of production we have always replaced labor with capital.
- The labor that is replaced by capital is not only unskilled labor, but skilled labor too. Under our system it makes no economic sense to spend a lot of money on new equipment and systems in order to replace cheap labor with expensive labor. In advanced capitalist societies the goal is to replace expensive labor with machinery and with as little labor as possible.[5]
- Whether employment increases or decreases after computerization, whether pay goes up or down, whether skills expand or wither are issues subject to constant negotiation on and off the shop floor, in high tech as well as mature industries. As a student of organizations and production, the most important differences for me between mature industries and the high-tech industries are not the technological differences—which are grossly exaggerated anyway (cf. Rumberger & Levin, 1984)—but differences of opportunity workers in mature industries have had to establish their claim to a slice of the pie.

[5] See Werner Wobbe-Ohlenburg's (1983) report on the introduction of robots in the Volkswagen plant in Wolfsburg, Germany.

- Whether people have interesting, challenging, rewarding, and skilled jobs is primarily a political issue, not an engineering problem. There is no quick technological fix in sight for boring, repetitive work, or, for that matter, for the problem of no work at all.

ACKNOWLEDGMENT

This research was sponsored by the Center for Prevention Research, NIMH.

7 Predicting the Use of Technology: The Case of Telework

Robert E. Kraut
Bell Communications Research

ABSTRACT

Telework is the use of computers and telecommunications equipment to do office work away from a central, conventional office. Today, a large number of jobs that are currently performed in conventional offices potentially could be performed at home. Both the jobs and the technology exist. Decisionmakers worry about the productivity and supervision of teleworking workers, but the available, sparse research suggests that the productivity of remote workers increases and that supervision can be satisfactory. Currently, however, the same technology that allows people to substitute work at home for work in a conventional office is used to reinforce conventional office work. People use it to supplement a conventional workplace by taking work home in an "electronic briefcase." Corporations use it to export conventional offices from central cities to suburbs and foreign countries with low labor costs. A major reason for this lack of workplace substitution is the potential threat to basic facets of organizational and personal life that work at home poses.

Recently a number of scholarly articles and articles in the popular press have argued that new computer and telecommunication technology is enabling flexibility in work arrangements that was previously unavailable (e.g., Becker & McClintock, 1981; Giuliano, 1981a & 1981b; If home is where the worker is, 1982; McClintock, 1981; Nilles, Carlson, Gray, & Hannerman, 1976; Olson, 1982, 1983; Schiff, 1983; Toffler, 1979). In the extreme, Toffler argues that new technology enables an "electronic cottage" and the radical transformation of work in America. In a more cautious and scholarly tone, Olson (1983) argues that the technology enables "space and time independence" of work. Commentators predict by the 1990s that between

10% (e.g., Kelly cited in Chabrow, 1985) and 40% (J. Carpenter, cited in Bell Communications Research News, 1985) of work will be performed from home using computers and telecommunications equipment. These predictions are based on changes in the occupational structure of the United States economy and changes in the technology with which people do work. But they also are based on a technologically deterministic view of the world that assumes if technology is appropriate it will be applied. By 1980, about 40% of the United States work force were office workers (Panko, 1984). Commentators make the claim that office automation technology—the electronic storage, retrieval, manipulation, and communication of records, messages, documents and other information—frees office workers from the confines of a standard 9-to-5 business day in a conventional office. Because white-collar workers can remotely receive, send, and act on the information they need to do their jobs, they can now work whenever and wherever their circumstances and their bosses allow them.

The intention of this chapter is to examine two components of these claims, that remote work is a transformation of the way work currently is done in America and that new technology is a cause or at least an enabler of that transformation. I treat telework as a workplace innovation and examine it in the context of what we already know about the spread of innovations across people and across organizations. My conclusion is that the new technology, indeed, has the potential for transforming work, but that it does not now serve that function and it may never do so. The independence of space enabled by the telephone in the early part of this century did not lead to the dissolution of the conventional office (Pool, 1977). So too the current technology, per se, is unlikely to lead to a dispersed work force.

At the present time, much white-collar work is done outside of the confines of the standard workplace and standard worktime (Giuliano, 1981a), much of it without the help of computers or new telecommunications technology. In addition, the new technology that has been introduced into the workplace is being used to maintain current work styles rather than transform them. I conclude from these observations that the use of new computer and communication technology is neither a necessary nor a sufficient condition for the work style flexibility implied in the remote work or electronic cottage concepts.

What is Telework?

Telework, remote work, or telecommuting[1] is the use of computers and

---

[1] *Telecommuting* (Nilles et al., 1976) is a frequently used term for this phenomenon, but the word is an awkward back formation from such terms as *telecommunication* and *television*. *Tele* means at a distance or from afar and thus telecommuting would be redundant, meaning commuting from afar (i.e., standard commuting), just as telecommunication is communication from afar. *Telework* (Giuliano, 1981a, 1981b), working from afar or at a distance, is the better term and will be used here.

telecommunications equipment to do office work away from a central, conventional office. Work can be performed at the alternate site full time or for several days a week. The alternate work site can be the employee's home, a satellite work center close to the worker's residence, or transient locations such as hotels or public transportation. Although working in homes, hotels, and planes frees workers from the constraints of a conventional office, work in a satellite work center simply moves a conventional office closer to home without changing the work style likely to be performed at that site. In this chapter I concentrate on telework at home.

Large numbers of workers are potentially able to work at home, although as Kraemer (1982) points out, reliable estimates of the numbers of jobs that are appropriate for telework or the numbers of workers who might ultimately adopt teleworking are not available. Some work requires that the worker be at a particular location (e.g., assembly-line work or work requiring face-to-face contact with others). Other work doesn't require a worker in a fixed location. This is especially true for "information workers," whose task is the production, collection, transformation, and distribution of information (Kraemer, 1982).

The United States economy consists of large numbers of information workers. Porat (1977), using detailed input/output analyses of the United States market, estimated that by 1975 information workers had accounted for over 50% of the United States labor force. Unfortunately, even among information workers, we have no precise definition of the type of work that could be done remotely and no good mappings of our vague definitions to the U.S. Census Bureau's or Labor Department's occupational classifications. Some white-collar information workers may be able to perform much but not necessarily all of their work remotely. This would include such job categories as accountants, computer specialists, librarians, writers, editors, purchasing agents, sales managers, stock and bond sales agents, billing clerks, and stenographers. Other white-collar information workers, such as therapists, teachers, photographers, curators, insurance adjusters, office managers, bank tellers, cashiers, and mail carriers, could perform little if any of their work remotely as it is now constituted with current technology. A rough estimate of the maximum number of workers who could work remotely at least part time is 50% of white-collar workers (Harkness, 1977) or roughly 26 million workers.[2]

[2] Many professional occupations that are candidates for teleworking are growing quickly in percentage terms (e.g., para-legal personnel, computer systems analysts, computer programmers, and aero-astronautic engineers all have estimated growth rates over 70% from 1978 to 1990; Carey, 1982). The candidate occupations with the greatest growth in terms of numbers, however, tend to be clerical. For example, general office clerks, secretaries, typists, accountants and auditors, and hand bookkeepers all have projected employment growth of over 200,000 from 1978 to 1990 (Carey, 1981).

TELEWORK NOW

Energy conservation was the initial justification for research on telework (Harkness, 1977; Nilles et al., 1976). As a result, the primary focus of research on this topic has been on workplace substitution, that is, working at home or other remote sites in lieu of work at the central office. The thesis of this section is that remote work occurs to a large degree now, but in ways that do little to shake the dominance of work at a conventional office. Instead, individuals and corporations use computers and telecommunications equipment for workplace augmentation, in which work at home or another remote site is added to work done in a convention office. In addition, corporations have suburbanized work, moving components of their conventional offices away from central cities to the suburbs, and have exported office jobs as well.

National Samples

Despite optimistic predictions, workers who work at home represent only a small proportion of the nation's workforce. According to Census Bureau figures, the proportion of people who work at home has been decreasing. The decennial census asks respondents about their means of transportation to work during the week preceding the census survey. Respondents could indicate that they worked at home. This census question is certainly not an ideal measure of the prevalence of home work in general and of telework (i.e., electronic home work) in particular. For example, respondents were asked only about their primary work site. Multiple job holders were asked only about the job at which they spent most hours. The census provides no information about the technology used on jobs. The target week may not have been a typical work week. And yet the census is the only source of data on a large, representative sample of the United States labor force and is therefore the source of the best available data in the United States on the prevalence and distribution of home work.

In 1960, only 3.6% of the nonfarm labor force worked at home as their primary place of employment. In 1970 the figure was 2.0%, and in 1980 it was 1.6% (unpublished data computed from the U.S. Census public use micro sample). These figures are, of course, old and do not encompass the 1980s when personal computers were spreading rapidly through corporations and homes. They do, however, encompass the last major wave of office automation, in banking and insurance during the 1970s.

Table 7.1 shows the distribution of home work among occupations. We see a small amount of home work among white-collar workers: executives, professionals, technicians, sales, and administrative support. In 1980, if we look only among white-collar workers, home workers were disproportionately managers and executives, writers, editors, and entertainers, sales

TABLE 7.1
Percentage of Home Work by Occupation

Occupational grouping	*N (thousands)*	*% home work*
All nonfarm workers	82,235	1.6%
Executive, administrative, & managerial	10,079	2.1
Professional	12,409	1.9
Technicians & related support	3,002	.6
Sales Occupation	9,478	1.8
Administrative support, including clerical	16,441	1.1
Private household service	480	11.7
Protective service	1,276	.4
Service, except private & protective	10,089	2.3
Farming, forestry, & fishing	1,552	8.8
Precision production, craft, & repair	12,396	1.1
Machine operators	6,156	.6
Fabricators, assemblers, & hand work	3,073	.6
Transportation & material moving	3,336	.5
Handlers, cleaners, helpers, laborers	5,385	.5

Note. Computed from U.S. Bureau of The Census (1980).

people, information clerks, social workers, and social scientists. On the other hand, white-collar workers who disproportionately work in conventional locations include communications equipment operators, college teachers, scientists and engineers, mathematics professionals, health professionals, and administrative support personnel.

These data, of course, estimate the amount of work done from home regardless of technology. Thus, daycare providers, who run a daycare business in their basements are counted as well as authors who write on a personal computer or computer programmers who communicate with a mainframe computer in another state. If we restrict our search to telework—the use of computers and telecommunications to substitute for travel to a conventional work site—the data are virtually nonexistent and the estimates, unreliable as they are, are much lower. For example, Electronic Services Unlimited (1984), a consulting firm specializing in telework, after reviewing the literature and analyzing a number of company pilot projects, estimated that there were about 10,000 teleworkers in the United States.

Although few people work at home as their major place of employment, many people work at home as an adjunct to their conventional office jobs. An AT&T (1982) national probability sample of households found that a large proportion of American households had members who worked at home on at least a part-time basis. Of the households in the AT&T sample, 54% contained individuals who were employed outside the home; more than 30%

of those employed brought work home with them. That is, about 16% of the households in their sample had workers who brought work home from outside employment. Those who brought work home were primarily employed in white-collar, information sector, office jobs. Thus, 88% of those who brought work home were in managerial, professional, technical, sales, or administrative support occupations compared to only 52% of those who brought no work home. Similarly, 60% of those who brought work home had an office or school as the location of their outside employment versus only 36% of those who brought no work home. These home workers tended to work in the evenings and on weekends.

In addition, 22% of households had some income-producing business conducted from home, generally small (grossing less than $5,000), young businesses (less than 2 years old) employing only the proprietor. These tended to be service or product sales businesses.

The tasks that both groups—those who brought work home and those who operated a business from home—performed at home did not require sophisticated computer or telecommunications equipment. They used the telephone frequently. Over half also read and did clerical and financial work, although financial work was not done frequently. Besides the telephone, paper, pencil and calculators were the only frequently employed tools.

In summary, these national samples show only a small proportion of workers whose home is their major place of employment. Furthermore, this proportion is decreasing. On the other hand, the data show large numbers of information workers who occasionally bring work home, using the telephone, but no sophisticated computer equipment. In addition, a surprisingly large number of small sales and service businesses are run from home.

High Technology Case Studies

We can get another perspective on the numbers of information workers who use telecommunications technology to work at home by looking at the behavior of workers in high technology companies. (See also Olson, this volume). These workers have relatively sophisticated computer and telecommunication technology available to them in their conventional office work, may have similar, company-provided facilities available at home, and are likely to be among the American households who currently have personal computers at home. They are therefore likely to use or want to use the technology available in the conventional workplace when they work at home. As such, they act as innovation leaders (Hippel, 1983), previewing the behavioral adaptations that the general public is likely to make as new technology becomes more widely distributed.

IBM. As an example, consider IBM's telecommuting experiment (Miller, 1983). In 1981, because the management at IBM feared future shortages of computer scientists and electrical engineers, they conducted a large scale experiment in which 300 volunteer programmers worked at home. The management believed that teleworking would allow them to attract and retain qualified workers. IBM implemented teleworking on a larger scale for programmers and marketing representatives after establishing through their trial that employee productivity had increased and that the company could manage logistics and security. In 1983, IBM paid for telephone line, modem, and terminal costs for over 8,000 professionals who worked at home. Company guidelines stated, however, that the home work was not in lieu of work during regular hours at a company location. According to official company policy, these employees could work as much as they wanted at home as long as they put in a standard work week in the office. (Informal arrangements, of course, may have been different.)

Other Teleworking Experiments. The National Academy of Sciences (1985) recently held a symposium on telework, at which a number of organizations described their experiences with this form of work reorganization. Five large employers described pilot projects they had initiated with teleworking. These projects typically allowed a small number of highly motivated, volunteer employees to perform most or all of their normal work from home. I briefly describe the organizations' evaluations in the following section on benefits. For the present purposes, I should just note that most organizations were satisfied with their ability to manage employees who worked from home and with these employees' productivity. However, perhaps the most surprising finding reported at the National Academy of Sciences' conference was that despite the testimonials these organizations bestowed upon their telework experiments, none was dramatically increasing the numbers of its teleworking employees to even 1% of its work force. For none of these organizations was telework to be routine.

Informal Work at Home. In 1983 and 1984 I conducted a study of over 300 computer users at AT&T Bell Laboratories and AT&T Information Systems to examine informal home work arrangements in a high technology company. (Additional details of the research are available in Kraut, in press.) Respondents were employed by research and development companies that had, to a large extent, automated clerical work and provided managers and professionals with relatively mature computer-based tools to perform work. For example, virtually all document preparation in these companies was done using computer terminals attached to mini-computer-based text-processing systems. Much of the clerical work was done personally by professionals and managers, rather than through a clerical support staff. Virtually no professional or managerial workers had typewriters and almost all had terminals at their desks, ranging from stand-alone graphics workstations to personal computers, to dumb terminals connected to mini-computer hosts. Most technical employees were on corporate electronic mail networks that allowed them to communicate with and send documents to co-workers in their own departments or across the country; in addition, electronic mail gateways allowed them to communicate with researchers in universities and private research organizations. Electronic bulletin-boards and news programs brought in information from around the country. With the pervasive use of office automation technology in this setting, we would expect that workplace effects that might follow the widespread diffusion of office automation more generally would already have occurred here.

I sent an electronic survey to 727 computer users, selected at random from among the recent users of 14 company-owned computers in New Jersey, Illinois, and Colorado. Four hundred and fourteen users answered the questionnaire, of whom 327 gave relatively complete and usable answers.

Table 7.2 shows some characteristics of the sample. The sample was 68% male, 64% were married or living with a person of the opposite sex, had an average age of 32 years, and had an average of 17 years of education (i.e.,

TABLE 7.2
A Comparison of Office Workers versus Home Workers

Variable	*Office workers*	*Home workers*	*t (Office > Home)*
	($n = 169$)	($n = 153$)	
Age in years	31.7	33.7	−1.98
% female	30.3	24.3	1.19
Years of education	16.2	17.3	−5.44
% living with spouse	58.6	69.9	−2.13
Number of children	.49	.54	−.55
Minutes commuting	29.3	28.4	.52
Salary level (1 = low, 7 = high)	4.3	4.9	−4.4
% with clerical job	10	2	2.8
% with administrative job	6	6	.01
% with technical job	81	90	−2.17
Job autonomy (1 = low, 5 = high)	3.5	3.7	−1.6
Job satisfaction (1 = low, 5 = high)	3.0	3.1	−1.1
Work group satisfaction (1 = low, 5 = high)	3.7	3.5	−1.9
Hours worked at office	33.5	36.5	−3.4
Hours worked at home	0	7.5	−17.0
Hours not worked	.55	1.4	−3.4
Hours interested in home work	13.6	17.7	−3.4
Hours job would allow home work	16.6	19.6	−2.4

some graduate training). Most were technical professionals divided among researchers in mathematics, computer science, and the behavioral sciences, software developers, hardware developers, and systems engineers; the proportion of clerical workers and administrative workers underestimated the proportions working in the company as a whole.

The central concern of this research was identifying the amount of time workers actually worked at home and the factors that allowed them and impelled them to work there. It is important to note that according to company policy, a standard work week was 37.5 hours long and was performed on site. These was no company-wide provision for home work and no company-sponsored experiments on this work arrangement, but individual workers could work out informal arrangements.

The first major point to be noted from this research is that although many people worked at home on an informal basis, they tended to augment work at their conventional offices, not substitute for it. Although on average, workers worked a full work week in the office, about 50% of them also worked at home. Although some employees substituted work at home for work in the office on an hour by hour basis, others added time in the office whenever they worked at home. Overall, the more time people spend working at home the more time they also spend working in the office (or vice versa). Of the 322 respondents who answered the appropriate question, 153 had worked at home in the week preceding the survey, an average of 7.5 hours, (i.e., a full working day). But they also worked 36.5 hours at a company location, significantly more than the 33.5 hours put in by those people who did no work at home.

Second, employees would like to work at home much more than they did. Figure 7.1 shows this dramatically. The employees who worked at home an average of 7.5 hours per week were interested in working at home almost 18 hours per week. Those who at the time of the survey did no work at home were interested in working there almost 14 hours per week. Respondents to this survey also believed that their jobs would allow them to work 18 hours per week at home on average. Figure 7.1 compares the distribution of actual hours worked at home with the distribution of the number of hours employees wanted to work at home and thought that their jobs would allow them to do so. The discrepancy between these two distributions suggests that organizational constraints rather than personal or job-related ones had limited the amount that employees worked at home.

Third, employees who worked at home showed a division of labor across space. A comparison of the work they did at home with the work they did in their offices is presented in Table 7.3. Respondents estimated the frequency in the previous week with which they engaged in each of a number of work tasks, separately for tasks performed at home and in a company work location. One can summarize the findings by noting that employees use their conventional office for social tasks and their home for cognitive tasks. Thus,

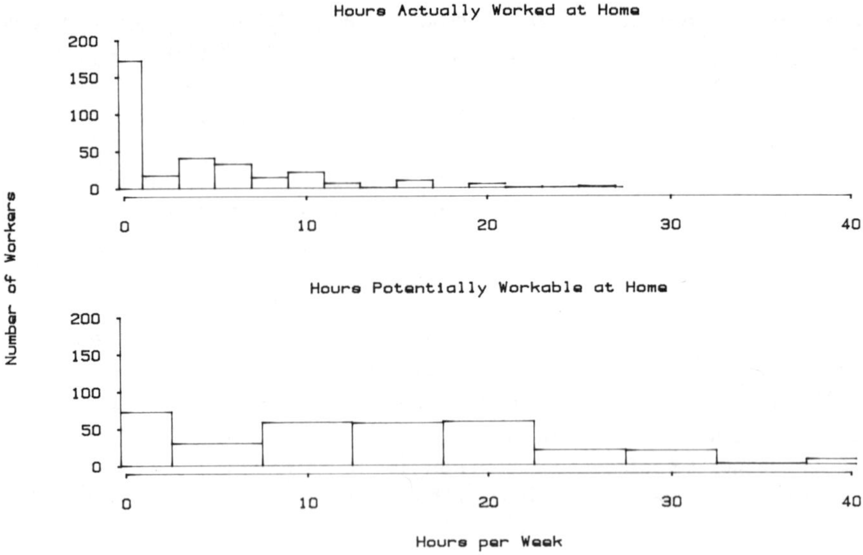

FIGURE 7.1. Comparison of actual versus potential home-based work.

as Table 7.3 shows, compared to their work in an office, when these employees worked at home they were more likely to read, write, and program, and less likely to talk to colleagues face-to-face or on the phone. Among the social activities, only the use of electronic mail was performed from home more than from the office. Presumably, this results from the asynchronous nature of electronic mail communication. Although telephone

TABLE 7.3

Tasks in Office and at Home

Category		***Office***	***Home***	***t* (Office > Home)**
Cognitive	Read	1.02	1.50	−5.0
	Write	0.78	1.16	−4.6
	Program	0.28	0.86	−22.1
Social	Talk about work (face-to-face)	1.30	0.16	1.5
	Talk about nonwork (face-to-face)	0.18	0.08	7.3
	Attend meetings	0.28	0.0	3.9
	Telephone	0.72	0.39	3.9
	Electronic mail	0.34	0.67	−4.3

Note. Entries have been standardized within location, so that a 0 indicates an activity was never performed at a given location and a 1 indicates that an activity was performed a standard deviation more frequently than other activities at that location.

communication can overcome the spatial isolation of working at home, it cannot overcome the temporal isolation.

Finally, the variables predicting the amount people actually worked at home were not the same ones that predicted the amount they wanted to work at home or the degree to which their jobs allowed them to work successfully at home. The effects of family life on working at home were complex. Interestingly, in this sample of salaried employees who performed supplementary home work, having a spouse led them to work at home, but having children made them less likely to want to work at home and to believe that they could do it successfully.

BENEFITS OF REMOTE WORK

Given that large numbers of people could potentially work at home, how many will ultimately do so and who will they be? It's hard to predict the future and harder yet when the past is so poorly understood. Our abilities to predict the extent to which telework will transform the nature of work in America must of course be based on our knowledge of current trends in the workplace, such as I've presented here. But, in addition, we can draw inferences from parallel processes, the spread of other technological and behavioral innovations.

The literature on the spread of innovations is a rich one, dating from anthropological and sociological work in the early part of this century (see Katz, Levin, & Hamilton, 1963; Rogers, 1983 for reviews). By 1983, the literature contained over 2,200 empirical books and articles as well as many more essayistic discussions. Despite the limitations of the work to date (see Brewer, 1980 for a good review), work on diffusion of innovation generally can raise our sensitivity to factors that may be important in predicting the spread of telework, as well as other workstyle innovations (cf. Walton 1975, 1977). A fundamental dictum in this literature is that characteristics of an innovation are only some of the many factors that influence its spread.

To understand the diffusion process, we need to know much about the potential adopters, especially their values and pre-existing behavior patterns, and the benefits they may derive from the innovation, because these are the attributes that have the strongest associations with the adoption of innovation across studies (Tornatzky & Klein, 1982). The following sections below consider in detail two of these characteristics: the potential benefits that adopters of telework are likely to derive and the behavioral and value shifts they are likely to endure.

Telework: Who Benefits?

Of what value is teleworking to its adopters? For telework to succeed both organizations as employers and the people they employ need to accept it. The benefits of telework, therefore, depend on whether we are considering

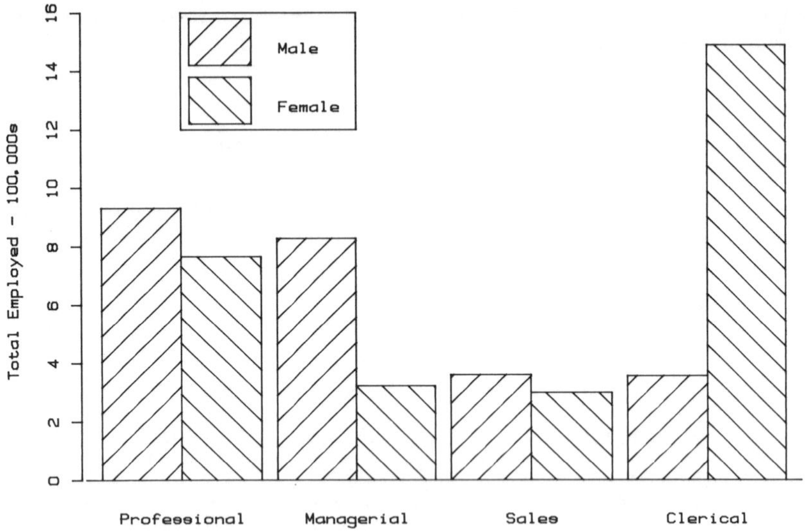

FIGURE 7.2. Occupation and gender of white-collar workers, 1982.

employers or employees. In addition, we must distinguish between professional/managerial and clerical/support workers.

The Professional-Clerical Distinction. Differences in the nature of the work of professional and managerial workers and clerical and support workers, differences in their labor supply, differences in their labor market characteristics, differences in their working conditions, and differences in their power all have implications for the suitability of telework for these workers and the benefits that they and their employers could derive from it.

First, consider the major difference in their sex composition and associated characteristics. As Fig. 7.2 shows, compared to managerial or professional work, clerical work is women's work. Overall, in 1982 the United States labor force was 43.5% female. Among professional and technical workers, 45.1% of the workers were female and among managers and administrators it was 28.0% female. On the other hand, clerical work is 80.7% female.[3]

---

[3] These summary statistics misrepresent the degree of occupational segregation by sex by aggregating heterogeneous collections of occupations. Occupational segregation by sex within particular occupations is much higher (Reskin, 1985) and within jobs in particular firms is almost total (Bielby & Baron, 1985). For example, although professional and technical workers are 43.5% female, some occupations suitable for telework within this class are heavily male (e.g., engineers are 94.3% male and lawyers and judges are 84.6% male) whereas others are heavily female (e.g., librarians, archivists, and curators are 80.7% female). Even among heavily female clerical workers, some occupations are heavily male (e.g., shipping and receiving clerks are 75.2% male) whereas others are almost exclusively female (e.g., secretaries are 99.2% female, typists 96.6%, receptionists 97.5%).

Women are an increasing proportion of the labor force (up from 33% in 1960, to over 43% in 1983). In 1960 35% of women were in the labor force; by 1980 this had risen to 52%. Much of this rise came from married women, especially those with children. In 1960 18.6% of married women with children under 6 in intact families were in the labor force; by 1981 this had risen to 48.7%. Most separated and divorced women with children have always had to work to support their children. As a result, their labor force participation isn't increasing as dramatically, because it started at a high level.

Given that clerical work is primarily women's work, that in our society women are primarily responsible for childcare, and that large numbers of women with young children work, a major difference between clerical workers and others is the extent to which they must integrate their work lives with their child-care responsibilities.

The second major difference between clerical workers and professional or managerial workers inheres in the work itself. Professional or managerial work is more autonomous than is clerical work (Hackman & Oldham, 1980). The greater freedom for professional employees to decide what tasks to perform and when to perform them, their greater freedom from supervision, their vaguer performance standards, and the longer deadlines associated with their tasks may make their jobs more appropriate than clerical ones for telework. Certainly, these factors change the style of remote supervision that would be appropriate for remote professional as opposed to remote clerical workers.

The third major difference between clerical work and professional or managerial work is in its legal definition. Clerical work is generally covered under the Fair Labor Standards Act. This act, as amended, covers such conditions of work as minimum wages and overtime pay. Professions and managers are generally exempt from these standards.

Finally, as Iacono and Kling (this volume) point out, clerical workers are generally at the bottom of an office power hierarchy and have little power to shape their conditions of work.

Benefits to Employers

Availability. The major benefits of telework to employers are personnel-related. Employers who allow or encourage their professional or managerial employees to telework generally expect two gains: a larger pool of scarce, skilled labor and increased productivity from that labor (National Academy of Sciences, 1985). Professional workers who might be more available as a result of telework are those who need to mesh family and work obligations (e.g., female heads of households or members of dual career families).

Employers of clerical workers often are interested in cheaper, more docile labor, although more abundant supplies of more productive labor also would be appreciated. In part, they achieve this cost reduction by coupling home work with an arms-length employment relationship to their employees, for example, by hiring them as independent contractors rather than as regular employees. Thus, employers of teleworking clerical workers typically pay piece-rate wages to temporary workers who receive no fringe benefits (e.g., Geisler, 1985; Olson, this volume).

Productivity. The data on the effects of teleworking on workers' productivity are sparce; they suggest that people who work at home are more productive than office-based workers. McClintock (1981), for example, interviewed 20 university professionals. He noted that increased time spent teleworking was associated with greater efficiency on routine work tasks, greater effectiveness on complex and nonroutine work tasks, and resultant gains in productivity.

Supplementing formal research by academics, corporations have evaluated their own pilot projects less formally. Teleworking has been tried in a small number of well-publicized pilot projects. Electronic Services Unlimited (1984), a private consulting firm, had identified 24 teleworking experiments or pilot projects with evaluation. The pilot projects typically had 6 to 25 highly motivated, volunteer participants performing most or all of their normal work at home. The pilot projects reported at a recent National Academy of Sciences (1985) conference on teleworking were unanimous in concluding that teleworking increased productivity. Productivity among both clerical and managerial teleworkers typically increased about 15% to 25%, using a variety of productivity measures, both subjective and objective. For example, Blue Cross/Blue Shield of South Carolina reported more claims processed with fewer errors for clerical workers keying in transactions (Geisler, 1985); the U.S. Army reported programmers writing more lines of code (McDavid, 1985); Mountain Bell reported course writers writing more courses that helped students learn better (Phelps, 1985).

These same studies suggest that employees were generally happy working at home. (The unhappy workers in these pilot projects were able to return to conventional employment.) To the extent that teleworking increased workers' job satisfaction, it should also increase their time at work and their commitment to the job (i.e., decrease absences, tardiness, and turnover), for both professional and clerical workers. Job satisfaction, itself, however, has no effect on productivity, the quality or quantity of output per unit time worked (Locke, 1976).

One must hesitate about accepting these findings, however, given the generally poor research methodologies, non-comparable control groups, and selective samples used in these evaluations. A careful reading of the literature suggests that to the extent that productivity gains are associated with telework, they are probably the result of highly motivated, volunteer workers putting in more time on their jobs when they were working at home than when they were working in a conventional office, (i.e., a self-selection effect). The gains may not be sustained with more general use of teleworking by less motivated workers who have no choice but to work at home.

Overhead. Finally, most employers can reduce their overhead costs by encouraging teleworking, either by moving some employees to cheaper locations (e.g., satellite locations away from expensive, central cities) or by shifting the cost of office space, equipment, or utilities to the employee (e.g., through unreimbursed home offices). Whether employers actually reduce costs, of course, depends on the trade-off between reduced overhead, increased productivity, and increased communications costs.

Miscellaneous benefits to employers include decreases in relocation costs and decreases in disability payments when disabled workers can maintain employment at home. When the teleworkers use central computing resources, doing so from home tends to distribute the computer use throughout the day and night and effectively decreases the organization's needed computer capacity.

Benefits to Employees

Some benefits to employees accrue regardless of the type of employee. Thus, teleworkers reduce their commuting time (which in the United States averages about 1 hour per day round trip), reduce their incidental costs (e.g., transportation,[4] clothes, or restaurant costs), and increase their flexibility to coordinate their work with their personal schedules and responsibilities.

Other benefits depend on whether the worker is professional or clerical and male or female. For workers who also are responsible for child care—especially female clerical workers, whose wages are too low to afford full-time, paid child care—teleworking provides an opportunity to combine at least part-time work with childcare. Table 7.4 shows a demographic profile of men and women who were identified as working at home on the 1980 U.S. Census. It shows that compared to those who work in conventional places, a home worker is likely to be a married women in an intact family with a young child, who is self employed, and works part time. This constellation suggests that home work is indeed being used to combine paid employment and childcare (see Kraut & Grambsch, 1985). Similarly, tele-

---

[4] Reduction in transportation costs is in the national interest as well as in the interest of particular employees. The original research on teleworking (Nilles, et al., 1976) was interested in the cost trade-offs between travel and communication, especially as this trade-off affected energy needs. Nilles et al. predicted substantial declines in energy consumption with increases in the numbers of people working at or near their homes. Harkness (1977) estimated that 25% of all United States motor fuel consumption and 11% of the nation's petroleum consumption were due to commuting or the journey to work. For example, in 1980 the 38 standard metropolitan statistical areas with populations of 1 million or more represented about 49% of the employed civilian labor force. Approximately 79% of these workers commuted to work in private automobiles, 61.2% alone and 17.7% in car-pools (U.S. Bureau of the Census, 1983).

TABLE 7.4
Descriptive Statistics for Home Workers versus Conventional Workers (white-collar workers only)

	Work-site last week			
	On-site		Home	
Variable	*Male*	*Female*	*Male*	*Female*
N (thousands)	23,648	26,936	371.2	456.0
White	91.1%	87.9	95.6	96.1
Age	38.8	36.1	46.8	43.3
Years of education	14.8	13.5	14.6	13.5
Married couple household	57.1%	31.9	44.9	54.0
Own children at home	92.2%	85.0	66.5	95.2
Preschool children at home	17.9%	13.2	10.8	19.7
Work-limiting disability	4.9%	3.2	10.6	7.1
Part-year or parttime	25.8%	51.8	40.6	74.0
Self-employed or employee of own corp	14.0%	3.7	65.0	50.8
Wages & self-employment income (all white-collar workers)	$18.0K	8.0	12.0	3.0
Wages & self-employment income (full-time white-collar workers)	$20.0K	10.8	17.0	8.1
Below poverty cutoff	2.6%	4.1	8.3	6.7

Note. Computed from U.S. Bureau of The Census (1980).

working provides the handicapped, the elderly, or otherwise homebound workers with new employment opportunities.

Professional workers, with others to care for children (e.g., paid child-care or compliant spouses), gain a more distraction-free work environment that meshes better with their lifestyles. On the other hand, because women with young children and other homebound workers have few other options in the labor market, their situation leaves them vulnerable to exploitation by employers.

Organized labor is very sensitive to the exploitation that teleworking may invite. Labor unionists point to the abuses of the past in the garment, glove, and jewelry trades that led to the Federal Government's banning of some home-production work (Chamot, this volume; Chamot & Zalusky, 1985). These abuses included low wages, poor benefits, long hours without overtime pay, hazardous working conditions, and the use of child labor. These are the issues about home work that have concerned social reformers since at least the turn of the century (e.g., Cadbury, Matheson, & Shann, 1906). Similar abuses are starting to reappear in home production work (Johnson, 1982). In addition, the effects of merging home and work domains on the family, especially on children, are not understood. Organized labor argues that protective labor laws such as the Fair Labor Standards Act, which were instituted to curb child labor and other sweat-shop abuses, are unenforceable when employees work at home. As a result, the AFL-CIO and the National Association of Working Women (9 to 5) has called on the U.S. Labor Department to institute a ban on home clerical work (National Academy of Sciences, 1985, Appendix C).

DISCUSSION

The major question that remains from this review of the motivations for working at home and evidence on its current status is why there is so little true workplace substitution. We have seen that large numbers of people could potentially work at home; the nature of the occupations, of the work tasks, and of the technology all allow this. The analysis of the relative advantages of telework suggests that it would be useful to corporations and to managerial and professional workers. It may provide a work option for homebound workers and those with child-care responsibilities, although it may be susceptible to abuse. Yet, corporations who test teleworking on a trial basis rarely implement it on a large scale, even when they have deemed the trial a success. Workers in a high technology company who currently work the equivalent of 1 day a week at home do so relatively independently of their full-time work in a conventional office. For these workers there is a large discrepancy between the amount of time they actually work at home and the amount of time they would like to work at home and think that they could. Even people who do no work at home would like to work at home almost two days a week and believe that the nature of their jobs would allow them to do so.

It is of course possible that we currently see such little workplace substitution because we are only at the beginning of a social phenomenon that is accelerating. The diffusion of innovation literature shows that the interval from the introduction of an innovation to its ultimate acceptance by most of the relevant consumers can vary widely, from less than 10 years, in the case of some consumer products like the television, to almost a century, in the case of other products, like the telephone. More importantly, from the shape of the S-shaped curve showing the degree of acceptance of an innovation over time, one cannot tell during the beginning of the diffusion process what the ultimate acceptance of the innovation will be. In the beginning, innovation failures and innovation successes look much the same.

In the beginning of the innovation cycle, temporary circumstances may hinder the spread of the innovation. These include phenomena common to most innovations, such as size limitations on the formal and informal communication networks through which knowledge of the innovation spreads or high price, which decreases with increased production. The hindering phenomena also may be specific to the innovation. In the case of telework, for example, acceptance may be retarded until the majority of the information that workers use is available on-line or until the majority of co-workers are accessible through electronic mail. Similarly it is possible that computer and communication services need to be easier to use or that communications costs must fall or that communications speed and quality must increase before telework will become a wide-scale social phenomenon. We could expect, however, that these inhibitors will weaken over time, as the community of teleworkers increases and as proponents, change agents, and entrepreneurs work to solve technical problems.

The Stability of the Conventional Office

But we should not expect a widespread adoption of telework simply because the technology allows it to happen. A more resistant barrier to the spread of telework is its incompatibility with the current work ethos. Although the adoption of any innovation requires some change on the part of the adopters, the adoption of telework requires unusually large changes in an especially central and stable component of the contemporary work culture. The conventional office structure, in which office workers congregate in a central office, is an indirect outgrowth of industrial automation. By the early part of the 20th century, manufacturing had primarily changed from an home-based, artisan production process to a factory system, in which masses of workers were brought together to accomplish related tasks. This industrial model eventually permeated education and office institutions as well (Zuboff, 1983).

Office structure has remained virtually unchanged since the late 19th century, despite both major changes in office technology (e.g., account books, shorthand transcription, telephones, dictation equipment, adding machines, letter presses, mimeo-machines, typewriters, and electronic copiers). The office also has remained stable in the face of explicit attempts to reorganize it (e.g., job redesign, flextime, and participatory management). Even the major change in the composition of the office—the feminization of office work—did little to change the structure of the office (cf. Baker, 1973; Davies, 1983; Scott, 1982).

The introduction of the telephone to the office is an especially apt comparison to the current introduction of computer and high-speed telecommunications technology. The telephone in the early part of this century could have allowed the dispersion of the white-collar work force, but the evidence is that it had the opposite effect (Pool, 1977). Physicians were able to move their surgeries away from their homes, and the management of industries moved from factories to conventional offices in central cities, not to their homes.

What causes this stability in office structure? My conclusion is that conventional, 9-to-5 office arrangements support a large number of activities crucial to the functioning of any work organization. Radical changes in the conventional office have the potential to disrupt these other activities and may be resisted.

The defining component of the conventional office is the co-presence of other workers for substantial parts of the work day. The presence of co-workers is the basis for a major satisfaction that workers derive from work: socializing and friendly social interaction (Locke, 1976). It also is a source of the social support that, among other factors, makes employment a contributor to the mental health of workers (Thoite, 1983). Social interaction as a source of satisfaction and support is even more important for office work than for factory work, because much of the work itself involves face-to-face dealings with others (e.g., Bair, this volume; Teger, 1983).

The presence of co-workers also facilitates the socialization of new workers to the workplace. The literature on socialization among both children and adults shows that modeling and observational learning are major ways for people to learn particular skills and general attitudes and values. In a modern office, the skills may be very specific ones, such as how to operate the copying machine or the text editor or how to complete an expense voucher, or they might be general ones, such as how to write well, to be a "good worker," or to get ahead.

In a related vein, the presence of co-workers aids the informal communication networks through which much information in an office flows. Informal communication networks generally follow proximity (e.g., Festinger, Schachter, & Back, 1950). People communicate most with, and are most influenced by those with whom it is easiest to communicate, those literally next door. The informal communication networks aid in the flow of both organizational norms and organizational information. Organizational norms include, among others, standards for dress, punctuality, and productivity. Organizational information includes gossip about individuals through which some organizational norms are enforced, information about organizational goals, and information about organizational structure (e.g., transfers and job openings). In addition, the informal communication networks aid organizational creativity, by fostering communication and collaboration among individuals who happen to be near each other.

The conventional office aids in at least one style of supervision or staff management. Managers can use the informal communication network to communicate tasks and goals and can get feedback about a subordinate's performance both from the informal communication network and from direct observation.

Finally, the conventional office structure helps workers structure their time. A basic social arrangement in modern society is that the individual tends to sleep, play, and work in different places, with different co-participants, under different authorities, and without an overall rational plan (Goffman, 1961). One of the ways individuals have adapted to the consuming and often incompatible demands of work and family life is to spatially and temporally segregate them. Telework by definition breaches this segregation and exposes workers to potential role conflict.

This discussion described several organizational and personal arrangements that are likely to be disrupted by telework. One might be able to devise technological and organizational solutions for some of the disruptions. The far-reaching nature of these disruptions, however, is an important reason why telework may not spread as a workplace innovation.

Alternate Routes to Organizational Goals

The second major reason why telework as a workplace innovation may not spread is that employers have alternative means through which they can reap the benefits of introducing new technology into office work. They can use the new computer and communications technology to automate office work and to spread conventional offices to areas of more abundant or cheaper labor. These alternate routes do not disrupt conventional office structure and the activities it supports.

First, employers can partially solve the problem of insufficient labor supply by using the new technology to replace workers, rather than using telework to attract them. Two technological trends are leading to increased replacement of human labor and mental activity with computers; both indirectly stem from continual decreases in computer costs and increases in computing and telecomunications capabilities. The first trend is the increasing availability of information in computer readable form. The information may have originated in this form, as when a customer orders from a mail order catalog by punching the buttons on a touch-tone telephone, replacing data entry clerks, or when an author writes a manuscript on a personal computer, replacing typesetters. In addition, the increased pattern recognition abilities of computers means that other information can be made computer readable without labor intensive transcribing of information by human operators. Thus, optical character recognition and voice recognition can be used, for example, to replace data-entry clerks in the insurance industry, one of the clerical jobs for which telework is currently being used.

The second technological trend is the increased abilities of computer software to incorporate human knowledge and intelligence. The intelligence of computers includes the ability to encode unambiguous knowledge and decision rules, as, for example, in accounting or statistical analysis programs or programs to compute a loan client's credit risk. It also includes the ability to encode ambiguous knowledge and heuristic reasoning processes, in, for example, medical diagnostic programs, equipment configuration programs, or computer program writing programs. This capturing of human intelligence decreases the demand for some skilled professionals, either by automating their functions, allowing less skilled para-professionals to perform some of the professional's tasks, or by making the remaining professionals more productive.

The second way in which employers can use new technology is to relocate conventional offices from areas with high costs and poor labor supplies to more favorable sites. The trends have been to relocate those parts of an office that require no face-to-face contact with the public (i.e., back office work) from northern central cities to southern and southwestern locations, to suburban locations, and to foreign locations. These sites have in common large numbers of potential clerical workers, generally nonunionized, who will work for below the prevailing wage rates in United States northern cities (Noyelle, 1984; Working Women, 1985). The important point to note for the present purposes is that the conventional office structure is being reproduced in these alternative locations.

We have seen that employers have alternate ways to meet their goals that do not entail the organizational and personal disruptions associated with telework. Given this state of affairs, if telework spreads as a workplace innovation, it is likely to happen because workers themselves find it in their interest to work at home. We have seen that managers and professionals seem to benefit most from telework. In addition, their jobs are likely to be most autonomous and therefore most suitable for telework. Finally, because their skills are in high demand, they are likely to be able to arrange the personal accommodations with their supervision to enable them to telework without wholesale changes in company policy. Thus, as a result of these considerations, if telework spreads, it is likely to do so as a bottom-up social movement spreading on an individual basis among managerial and professional workers through normal, individual diffusion processes.

SUMMARY

In summary, we have seen that telework is feasible and has benefits for employers and at least management and professional employees. It may not, however, spread as a workplace innovation because it threatens to disrupt too many basic facets of organizational and personal life and because employers have alternate ways to reap technological benefits without such widespread disruption.

8 Telework: Practical Experience and Future Prospects

Margrethe H. Olson
Graduate School of Business Administration
New York University

ABSTRACT

This chapter discusses current experience with work at home, primarily from an organizational perspective. First, the impact of technology on work organization is discussed, leading to the conclusion that the factors making work at home feasible are primarily factors of work organization. Four profiles of at-home employment that are based on two critical factors are described: the degree to which the employee's skill is in demand and the severity of the employee's "nonwork" constraints limiting employment choices.

The second part of the chapter describes an exploratory survey of 14 companies with formal work-at-home pilot programs and 6 companies with informal work-at-home arrangements. Although the pilots were viewed by management as moderately successful, most were not continued on a regular basis. The tentative conclusions from the survey are the following: (a) The primary interest in work at home of organizations is based on the need to acquire skills in short supply. The need for employees to have flexible work options is not a major concern. Traditional "on-site" management styles provide a major deterrent; (b) The technology supporting work at home is not a major factor today. In the future, when highly integrated office systems exist, technology may play a greater role; (c) In general, organizations will be altered through decentralization facilitated by location independence of jobs. New organizational forms may increase both organizational and individual flexibility.

The purpose of this chapter is two-fold: (a) to analyze telework from the point of view of work organization and personal choice; and (b) to provide a brief overview of company experiments with telework. The reader is referred to the previous chapter for a complete discussion of the social and organizational factors encouraging or inhibiting the growth of telework as an employee work option. The emphasis in this chapter is on work at home.

The first two sections provide two analytical frameworks for understanding the phenomenon of work at home. The first section describes the elements of work organization and argues that work at home may be considered an extension of certain forms of work organization rather than a separate phenomenon; many presumed advantages and disadvantages of work at home are actually results of work organization rather than work location. The next section categorizes the types of work at home that exist today, based on the author's assessment of current research, including her own, on the phenomenon.

In the remainder of the chapter, a particular type of work at home, a formal ongoing arrangement between an organization and an employee, is examined. The results of an exploratory survey of such programs by the author are presented. Finally, the author summarizes critical issues related to work at home that require further research.

TECHNOLOGY AND WORK ORGANIZATION

It is argued here that telework, particularly work in the home, is currently feasible only for certain types of jobs organized in a certain way. In fact, the capability for a job to be done in the home is basically an outgrowth of the way technology is applied to the organization of the work itself. To begin, then, a framework for analyzing the options in terms of work organization is presented. Although this framework is appropriate for any kind of work, the emphasis is on its application to "office" work.

The organization of work can be characterized by three factors: horizontal division of labor, vertical division of labor, and method and location of control.

Horizontal Division of Labor

Horizontal division of labor refers to the degree to which the total number of tasks are divided among individuals as well as to the complexity of those tasks. At the extreme of a high division of labor, the work is subdivided into the smallest units (or tasks) possible and a different person performs each unit repetitively. The goals are to improve the efficiency of the entire process and to reduce the skill level required of individual employees. The roots of this approach to the division of labor lie in Scientific Management, an approach to industrial work conceived by Frederick Taylor (Braverman, 1974). At the other extreme, often referred to as *job enrichment,* a single person performs many related tasks in an integrated fashion to produce a single whole unit of output.

Vertical Division of Labor

The vertical division of labor refers to the degree of separation of conception or planning aspects of work from its actual execution. With industrial work and Scientific Management, a significant degree of the mental work involving planning, design, calculation, and creative intellectual thinking is removed from the physical processes of carrying out the results of the mental effort. Furthermore, determination of production in terms of pacing, production expectations or quotas, and surveillance is concentrated in the work of the planners (Braverman, 1974).

With office work, vertical division of labor is exemplified by the degree to which the work of a so-called principal (who conceives of an idea and its manifestation in the form of a document, design, graphic, oral presentation, etc.) is separated from the work of a support person, whose task is to correctly or appropriately produce the manifestation of the principal's conception with minimal regard for the concept itself. With a low vertical division of labor, a single person with access to the necessary information and technology may execute both the conception and manifestation aspects of the work.

Control Over the Work

Control over the work performed by an individual can lie somewhere on a continuum from external control, where work pacing or outputs are dictated and control information is unavailable to the worker, to internal control, where the individual worker determines his or her own pace or output levels and control information is readily available for individual feedback. At the extreme of external control, the work may be machine-paced or output may be based on very specific short-term volume quotas; furthermore, the work can be monitored continuously and/or electronically, so that the employee is not aware of physical surveillance. At the extreme of internal control the employee has access to all the necessary resources (primarily information) to perform the task, has autonomy to set his or her own work pace and schedule of tasks (within some reasonable limits), and has complete access to control information as well as knowledge of how and when this information is being used by others.

Telework and Work Organization

Work at home may be considered simply an extension of the organization of white-collar work using office technology. Thus, it is a symptom or example of the alternative uses of technology rather than a separate phenomenon. Two examples illustrate this point: the first is a job with high horizontal and vertical division of labor and external control, whereas the second has low division of labor and internal control.

A good example of the first is an online order entry job, where a clerk copies data from a form to an online entry system via a terminal and exercises simple error correction procedures to handle discrepancies between the form and the online customer database. Different types of orders are separated into batches and each data entry operator "specializes" in a particular type of order only. Conception is completely separated from execution; the employee need have no knowledge of the ordering process or products to execute the required tasks. Production quotas as well as satisfactory error rates have been established; complete control information (number of documents processed and errors committed during any unit of time) are available to a supervisor via a specialized "monitor" workstation. Because the work is simple and repetitive, only basic typing skills are required; other training and the employee's general education are not important. Because the employee has no control over the pace or volume of work, there is little or no need for the employee to develop a relationship of trust or commitment with the supervisor. Criteria for employee selection can be low and in all likelihood there is little or no chance for promotion. Given these conditions of the work, it is only a simple step to shift the cost of space directly onto the employee by having the work performed at home.

An example at the other extreme of division of labor and control is a systems programmer. The tasks performed are highly complex and generally have a great deal of variety. Conception and execution are highly integrated in the process of, for instance, debugging a program on line. Furthermore, the employee has almost complete direct control over the work in terms of pacing, monitoring of outputs, and so forth. It is precisely because of the low division of labor and internal control that the work can be shifted easily into the home.

In the next section, a more complete categorization of the types of workers who might work at home and their conditions of work is presented.

PROFILES OF AT-HOME EMPLOYMENT

Table 8.1 classifies work at home on two dimensions: market demand for the employee's skill and the employee's reasons for choosing the arrangement. Under the latter, nonwork constraints refer to constraints such as child care,

TABLE 8.1
Types of At-Home Employment

Reasons for Choice	***Demand for Skill***	
	Low	***High***
Nonwork constant	Exploitation	Tradeoff
Personal choice	Autonomy	Privilege

physical disability, old age, and so on, which in a practical sense limit the employee's work options. Personal choice refers to a lack of those constraints; the employee chooses to work at home as a matter of personal preference or convenience. Whether skills are in high or low demand implies whether the person can bring in adequate income at home and/or has leverage with the employer. The terms used in Table 8.1 to refer to each profile should be taken as convenient labels for stereotypical cases rather than definitive terms. The profiles are described following.

Exploitation: Low Demand for Skill, Nonwork Constraints. If a potential employee has no particular skills that are in demand or is relatively unskilled in general, the available jobs will be limited. Those jobs that are available typically will be organized with high horizontal and vertical division of labor and external control, so that the skills required to perform the work are minimal. Thus, as argued previously, the job can be performed in the home without difficulty. Furthermore, if communications with a central site or computer are infrequent or sporadic, the cost of equipment to support the work at home is relatively low.

Furthermore, the potential employee is constrained in work choices because of nonwork responsibilities such as child care. Even if the employee could leave the home to work, the costs of commuting, child care, clothes, and so on associated with traveling to a central work site might be prohibitive. The potential employee might willingly accept a lower pay rate than would be available for work in a traditional location in order to gain the privilege to work at home.

There is a history of exploitation of workers in specific industries in this country, based on similar conditions to those just described, which have resulted in bans on home work in those industries; industrial sewing, knitting, glove making, and assembly of artificial flowers are examples. Recently, the federal ban on home knitting was revised; the current Republican administration favors a lifting of all bans on home work. At the same time, the AFL-CIO, one of the largest unions in the United States, has called for an extension of the ban to cover computer-based work precisely because of the perceived similarity to working conditions that led to exploitative use of home work in the past.

The term *exploitation* is a highly subjective one and needs some clarification. Many home workers who fall into this category (based on the nature of their work and their nonwork circumstances) have been interviewed by the author and others (Christensen, 1984; Olson, 1983). Overall, most of them have expressed a great deal of satisfaction with the arrangement because it permitted them to earn an income while caring for their children at home, something that to many was considered essential. Nevertheless, their work options were extremely limited and alternative methods of child care, if available at all, were often prohibitively expensive.

Autonomy: Low Demand for Skill, Personal Choice. The difference between this profile and the previous one is that the employee freely chooses to work at home even though other work arrangements are available. The stereotypical workers in this category are those whose needs for autonomy are high; they are therefore willing to give up higher or steady wages, job security, and opportunities for promotion, in order to "be their own boss." A number of books on work at home actually emphasize requirements for setting up one's own business (Applegath, 1981; Behr and Lazar, 1983; Edwards & Edwards, 1985; Hewes, 1981).

The stereotype of this profile is a self-employed person whose primary base of business is the home. The true extent of growth of this phenomenon is not well understood. Overall, the number of self-employed has been steadily decreasing (primarily due to the decrease in employment in the farm sector), but the number of self-employed in "white-collar" (non-craft, non-farm) occupations is not known.

Tradeoff: High Demand for Skill, Nonwork Constraints. Although all forms of work at home are in some sense a tradeoff for employee and employer, this profile represents the most distinct set of tradeoffs. The fact that the employee's skills are in demand means he or she has relative flexibility to move in or out of the workforce; however, many specialized skills, particularly those related to technology, need to be updated frequently. A typical example might be a professional who leaves the workforce temporarily for child rearing, but wants to keep his or her skills current in order to return to the full-time workforce after a certain period. The person may not be as concerned about the costs of going to work as the time to commute to and from a full-time job, even if high-quality daycare is available. Because the skills are in demand, the person can work flexible hours at home and take on some assignments and not others, depending on the dynamic circumstances of the person's "nonwork constraints."

Privilege: High Demand for Skills, Personal Choice. The idealized view of work at home focuses on this profile, a person who, because of the demand for his or her skills, has leverage with the employer. Because of personal preferences related to lifestyle, he or she may prefer to forego office politics and interruptions for working in the more relaxed environment of the home. The employer cannot afford to let the employee go and so tends to meet certain work "requirements" that would not normally be tolerated. A typical example is a highly paid systems programmer employed by a computer vendor. The job is low in horizontal and vertical specialization and control is internal; thus the job can be performed at home relatively easily as long as adequate equipment is provided. Obviously, few jobs or individuals actually fit this category.

AN EXPLORATORY SURVEY OF WORK-AT-HOME PROGRAMS

In this section, the results of a recent survey of formal company experiments with work at home are presented. They are contrasted with informal arrangements for employees to work at home that are purportedly common in "high-technology" industries, particularly in the "Silicon Valley" area of California. These arrangements are referred to as *informal programs* in the discussion that follows.

Formal Programs

Fourteen companies that have completed or are now conducting formal work-at-home pilot programs were surveyed. Nine companies had completed pilots and were able to report the results of their evaluations; two companies had only preliminary results to report, and the remaining three were just beginning their pilots at the time of the investigation.

The industries represented by participating companies are reflective of the types of companies that have expressed the greatest amount of interest to date in work at home. Nine companies represent the financial and insurance industries, which are heavily dependent on information processing and searching for ways to improve the productivity of this labor-intensive component of their work. Four are computer and communications firms that, while also seeking ways to improve productivity and quality of work life, are aware of the market that would be created for their products if work at home were to become a common phenomenon. One manufacturing firm (non-computer) was represented.

Reasons for Interest. Most of the reasons given by companies for interest in work at home reflect the long-term benefits that could be accrued if work at home were to prove feasible and were made available on an expanded basis, rather than actual benefits from the pilots themselves. Direct cost savings were seen for those pilots where a change in employee compensation resulted in reduction of employee benefits (see following). Increased productivity was generally recognized as a significant potential benefit even where it was not directly measurable.

The most common reason given was an interest in tapping an otherwise inaccessible labor force; this labor force includes those who require flexible work hours, such as women with full-time child-care responsibility, as well as those for whom commuting time and expense to work at a central office would be unreasonable. In other cases, work at home was viewed as one of a number of options for increasing employee flexibility, based on a long-term commitment to retaining valued employees. Two of the pilots specifically dealt with the severely (homebound) disabled, and were motivated by a corporate commitment to increase employment of the disabled. At least one company saw a significant advantage in shifting computer utilization to "off-peak" hours; this also was considered a productivity benefit because employees were able to receive immediate turnaround, allowing them to solve problems more efficiently. Those companies in the communications industry were implicitly or explicitly interested in the market potential for technology to support work in the home; management felt they should learn as much about work at home as possible and demonstrate its feasibility internally.

Job Descriptions. The types of jobs that were performed at home in the pilots are shown in Table 8.2. The term *computer professional* generally refers to programmers, programmer/analysts, or systems analysts. Although most of the pilots were aimed at a particular class of jobs (word processing or programming), four were open to any professional-level job that met certain characteristics and were approved by their management. (Two of these pilots included computer professionals.) Some of the other professional jobs were course development, internal auditing, and product management.

Employee Status. Table 8.3 shows the status of employees working at home in the organizations surveyed. This includes their employment status, the payment method, and whether or not benefits are received.

Equipment Support. Most of the pilots involved some form of computer and communications support in the home. There were two general types of configurations:

1. Terminal with telephone connection to the company's main computer. Employees typically also had a separate telephone installed for use with the terminal as well as a modem. In three cases special leased lines were installed, apparently for reasons of data security. A few employees utilized microcomputers rather than terminals, but this was not the norm.

TABLE 8.2
Job Classifications of Formal Work-at-home Pilots

Type of Job	*Number of Pilots*
Computer professional	5
General professional	4
Word processing	4
Clerical	1
Total	14

TABLE 8.3
Employment Status and Payment Method For Work-at-home Employees

Employment Status	*Payment Method*	*Benefits*	*No. of Pilots*	*Jobs*
Perm. full	Salary	Yes	7	Professional
			1	Word processing
Perm. part	Hourly	Prorated	1	Word processing
Perm. part	Hourly	No	1	Word processing
Perm. part	Output+	No	1	Clerical
Contract	Hourly	No	2	Professional
			1	Word processing*

+ Piece rates

* A temporary arrangement because of government requirements for the employees to be on contract for a temporary work period (disabled). They have since shifted to full-time salaried with full benefits

2. Word processor with dictation equipment. The four word processing pilots were set up for dictation transcription. All utilized stand-alone word processing equipment as well as equipment for transmitting and transcribing data and voice tapes. The range of sophistication of equipment support for these tasks varied widely. In one case, the employees also had a facsimile machine with a separate telephone line.

In most cases, the company paid for installation of equipment and monthly telephone charges. In one case the employees rented the equipment from the company.

Results and Current Status. Nine out of 13 companies were able to report on their evaluations of their pilot programs. Of these, six reported substantial to satisfactory improvements in short-run employee performance. In most cases these results were based on subjective evaluations by the employees and their supervisors; they found that they consistently met or exceeded deadlines for deliverables and were then able to take on additional work. In two cases of clerical work, significant improvements were reported in work turnaround and error rates.

The results of three pilots indicated some problems in terms of performance. These were generally attributed to lack of management commitment, communications problems, or technical problems.

According to management, employees were generally very positive about their work arrangements. It should be noted that in several cases, early participants expressed dissatisfaction with the arrangement and chose to return to the office full time. A more detailed discussion of employee reactions, based on interviews with employees themselves, is contained in a separate section of this chapter.

Reactions by management to the alternate work arrangements were mixed. Several pilots were terminated with a change of management. Managers of the word processing and clerical pilots were very positive about the programs. In several pilots involving professional employees, management expressed concern over the need for "remote supervision," even if they felt that employee performance was satisfactory.

Of the nine pilots that had been completed at the time of the survey, four were terminated and the employees returned to their former work arrangement full time. (In one of these cases some of the employees are currently working at home occasionally on an informal basis.) The remaining programs were retained or expanded somewhat and now continue on a regular basis.

Informal Programs

It should be noted that, despite the interest of the news media in work at home, few if any companies have adopted it formally as an employee work option. Although many companies are studying work at home, few have formal pilots and those that exist, like the examples given, usually involve only a few people. However, there also appears to be a trend toward more informal arrangements of work at home in certain companies in distinct industries, primarily the high technology computer industry located in California. A number of these companies were contacted in order to determine not only how prevalent work at home is, but whether the companies represent future trends in other firms. In particular, the role of technology was examined to determine its importance in facilitating work at home for a greater number of people.

The six companies participating represented three categories of high-technology firms: personal computer manufacturers, timesharing firms, and high-technology research and development firms.

The work-at-home arrangements in these companies were very similar. All were implicitly sanctioned, if not explicitly. All companies have relatively flexible work hours and make equipment readily available to employees for work at home at least outside of regular hours. The companies have a majority of professional-level employees and a relatively small clerical component. All employees who work at home are professional-level, with skills that are in demand. They are all on full salary and benefits; few work regular 40-hour weeks, and they may work substantial amounts of overtime on a sporadic basis.

The primary reason given by management for permitting work at home was that it helps attract and/or retain qualified personnel. The employees themselves tended to view it as a requirement for their continued association with the company (to be discussed following).

Equipment Support. Equipment support of employees, both in the office and at home, was usually substantial. In particular, those who performed software development for specialized applications required very specialized equipment in both locations. In some cases, the equipment at home consisted of lightweight portable personal computers or terminals, whereas in others those who worked at home had duplicate microcomputers.

Results and Current Status. For the most part, management reported that work at home improved employee performance, primarily because their work required intense concentration and their office environments did not provide it adequately. The employees themselves were very positive about it; many said they would leave the company if the privilege were terminated. For the most part, there was no negative reaction from other professional employees doing the same type of work, primarily because the option was available to them as well if they desired it. Supervisors' reactions were mixed but they tended to tolerate it. (More detailed statements about employees' reactions to the work at home situation are contained in the next section.)

EMPLOYEE REACTIONS: FORMAL VERSUS INFORMAL PROGRAMS

In order to understand better how employees themselves feel about working at home, 20 in-depth interviews were conducted: eight were participants in formal pilot programs and 12 worked at home under the "informal" arrangements. (Although many of the arrangements in the latter group were relatively formalized, they are referred to as *informal* in this section to differentiate them from the formal pilots.) In this section, the results of these interviews are presented with some comparisons between the two groups.

Employee Attitudes

Employees were asked whether work at home had a positive or negative effect on a number of aspects of their work and nonwork life. The results show some interesting contrasts between the formal and informal programs.

There was a distinct difference of reaction to supervision between formal and informal programs. Half of those in formal programs felt their relationships with their supervisors had worsened. Only one fourth of those interviewed in the informal programs felt the effect of work at home on supervision was negative.

The majority of employees felt the effect of work at home on their relationships with their coworkers was negative. This is not surprising because work at home must deprive them of considerable social interaction at work. Somewhat more surprising is the finding that at least some of the employees in the informal programs found that their relationships with coworkers improved. This is indicative of their preferences to avoid social interaction; they often commented that informal chats in the office distracted them from task accomplishment and that they would prefer to avoid them.

There was little evidence of decreased commitment to the organization as a result of work at home. Rather, almost half of those interviewed said that their commitment actually increased. It should be noted, however, that those in informal arrangements expressed relatively low commitment to the organization in general. Rather, commitment was strong to the particular product, project, or work group. Commitment to the organization was increased, if at all, because the employee was permitted to work according to his or her personal preference. Several joined the companies for which they worked specifically because they knew they would be able to work at home; several also said that if the company removed that privilege they might seek employment elsewhere. Where commitment increased for those in the formal pilots, it was generally because the company allowed the employees to resolve conflicts (i.e., family responsibilities) that would have made it difficult for them to work otherwise.

When asked about compensation, all but one of the word processing and clerical employees felt their overall compensation was reduced; the one employee who did not feel this to be the case was disabled and had not been able to work previously. These employees tended to have family responsibilities and work at home was one of few options available to them for bringing in additional income while still maintaining child care. They generally felt that, even if they could make more income doing the same work outside the home, they would incur additional costs for commuting and child care that might make up for the difference.

Overall, employees were very positive about the effect of work at home on their personal work effectiveness; they felt they produced better quality work at home whether or not it was recognized by management.

Generally, employees found working at home helped reduce stress. The lack of interruptions and avoidance of "office politics" were reasons given; another important factor is the elimination of stress associated with commuting. However, some employees found that work at home increased stress, primarily when they tried to juggle full-time job responsibilities with full-time family care. The sources of stress (whether job or family) were not apparent, but on at least this dimension it appears that work at home cannot be called a *good* solution to child care. Whether or not stress was increased, employees reported consistently that the work-at-home arrangement improved the quality of child care, primarily because they could do it themselves. Job satisfaction increased overall for both groups of employees.

In terms of the quality of employees' nonwork lives (besides children), the effect of work at home appeared to be very positive. Most reported a positive effect on both their time for leisure activities and the types of leisure activities in which they engaged. In particular, those in the informal group (most of whom did not have child-care responsibilities) took advantage of free time during the day to engage in sports such as skiing, hang gliding, windsurfing. Most of them live in California where such outdoor sports are readily available; several took advantage of the opportunity to engage in such sports and still avoid weekend crowds. Most employees with families also reported improved relationships with other family members.

It has sometimes been hypothesized that work at home may help to place a person's relationship to his or her community above their relationship to work; people may live in a community of their choosing rather than one that is proximal to work. Based on this small sample, however, there was little evidence that relationship to the community was altered by work at home. For the informal group, this may have been as much a function of the "California lifestyle," which is very individualistic and noncommunity-oriented, as of working at home. None of those in the formal programs became more involved in community activities either, and this could not be explained by regional differences.

Comparison to Profiles of At-Home Employment

Do these employees fit into the stereotypes described earlier? There is a risk of over-generalizing but at least two profiles are well represented. First, all 12 employees in the informal programs had skills that were in demand; furthermore, only 1 of the 12 had nonwork constraints (i.e., child-care responsibilities) which influenced the decision to work at home. Clearly then, these employees (with one possible exception) fit the "privilege" profile. This is supported by their preferences to avoid social interaction and their low commitment to the organization, as well as their reports of improved work effectiveness and improved family relationships. A tentative conclusion regarding the informal programs is that the difference between these and other firms is not the sophistication of their technology, but the skills and bargaining power of their employees. Thus it is not likely that the informal work-at-home arrangements represent future trends in other firms.

The second profile that is well represented by these employees is "tradeoff." For the most part, their skills (programming, word processing) were in short supply. However, they also had significant nonwork constraints (child care or disability) which foreclosed other work options. They missed the social interaction of the workplace and were concerned about their relationships with their supervisors. In general, they felt the arrangement had reduced their potential earnings. They also experienced stress in combining work and nonwork responsibilities. Overall, they viewed the arrangement as a tradeoff. There were certain disadvantages which were accepted because of limited alternatives.

SUMMARY AND CONCLUSIONS

In this section, the findings about work at home are summarized and interpreted. In order to accomplish this, work at home must be viewed from three different perspectives: social, technological, and structural.

Work At Home as a Social Phenomenon

It can be seen from the cases that the primary motivation of business organizations to investigate the option to work at home is economic. Companies are concerned with cost-effective ways to acquire skills that are in short supply. Quality of work life or the need to give employees flexible work options in order to accommodate their nonwork requirements (i.e., family), does not seem to be a major concern. Companies are interested in attracting a labor force that is not accessible under normal circumstances and are willing to relocate jobs in order to accomplish this. It also must be kept in mind that the majority of jobs for which work at home is considered (clerical, word processing and computing professionals) are not unionized.

Individual employees, on the other hand, are increasingly motivated to pressure their employers to permit more flexible work options, including work at home as well as job sharing, flextime, permanent part time, and extended leaves of absence. Their reasons for needing increased flexibility are clear: the increasing numbers of dual-job and single-parent families, the increasing number of mothers of small children who are entering the work force, and the lack of adequate day care facilities in the United States. Work at home is a relatively extreme, and far from ideal, solution to working and providing child care at the same time.

For some individuals, work at home is an ideal option because of personal preference. Some people prefer the informality and lack of disruptions that work at home provides; others prefer to live somewhere for lifestyle reasons that would preclude work for a company unless they work at home. So far, however, our evidence shows that only those individuals with skills in demand are in a position to request such special arrangements. Underlying these preferences is also often a desire for individual autonomy that overrides organizational commitment. Companies are seeking a desired skill and individuals are in a position to provide it; there is little sense of concern for quality of work life or long-term commitment to employees.

Work At Home as a Technological Phenomenon

Much of the interest in work at home in the United States has been motivated by the impression that new technological developments will permit more jobs to be conducted from remote sites; therefore, more people will be able to work at home. Based on our observations, we conclude that technology is a facilitator but not the driving force; strong organizational "cultural" factors help discourage it while individual needs outside of the work domain may encourage it. However, it is true that significant changes in technology may see more and more jobs performed "remotely" in some sense.

It has been predicted that with office automation technology, as many as 50% of all office jobs could be performed at or near an employee's home by 1990 (Harkness, 1977); the location of the jobs is not important and neither is the time in which they are performed (within limits, of course).

In most American business organizations today, the technological support that would facilitate work at home does not exist. The technology required has several components:

- Computing power that is inexpensive and portable.
- Access to all information resources required to perform the work in a form that is accessible by computer, requiring that it be both "machine-readable" and "online."
- Access to other people in the organization through communication networks that link all locations, office and home.

In the future, when highly integrated office systems containing these components exist, the technology may play a more significant role in providing needed flexibility for employees. We suspect that the most pervasive impact will be location independence, the general capability to be in touch with work from other locations and around the clock. Whether location independence is good or bad for quality of work life remains to be seen, and the phenomenon is in need of research.

Work At Home as a Structural Phenomenon

The third way that work at home can be viewed is as part of a more general structural phenomenon; the structures of organizations may be altered through decentralization facilitated by location independence of jobs. Basically, when integrated office systems are in place many more jobs may be performed remotely from a central location, and the need for central office structures is reduced. Companies in metropolitan areas such as New York City are moving their "back office" operations to suburban areas, primarily because of lower building costs and a qualified available labor pool. Other companies are moving certain operations to other states for tax purposes or moving work such as data entry "off-shore," to other countries where the price of minimally skilled labor is substantially lower. Technology facilitates these moves but is not the primary motivator.

In the future, companies may continue to become more physically decentralized as well as more specialized, continuing to look for ways to supply themselves with certain specialized skills at reasonable cost. The technology will continue to facilitate this trend as well as increased flexibility of organizational boundaries. Neighborhood work centers, where technical support (terminals and personal computers, teleconferencing facilities, printers, facsimile machines, etc.) is provided for employees from multiple organizations, may become a reality. In the United States, no experiments with neighborhood work centers have occurred to date.

The argument that telecommunications may be utilized to replace transportation and thus save energy may become a reality. One can envision community structures growing around a common lifestyle or recreational purpose rather than proximity to a central office location (the motivation for the development of suburbia). Such arguments have been made for a number of years and continue to be made; the realities of organizational structure and "culture," primarily in resisting change to old methods of working and managing, make such changes occur much more slowly in reality.

Informal Arrangements: The Organization of the Future?

The examination of a few selected high-technology companies was motivated by a desire to determine if they represent the typical company of the future, integrated through advanced technology and thus able to operate with location independence. Although these companies have an exceptional amount of technological support, our tentative conclusions were negative.

The four companies (not counting the two in research and development) are unusual in and of themselves. They are experiencing exceptionally fast growth in an extremely dynamic and high-risk industry. They are still, however, relatively small, with few levels of management and few signs (yet) of bureaucratic administration. There is a high component of professional (non-managerial) employees, and a relatively small component of clerical employees. Although there is an exceptional amount of investment in technological support per employee (required for the work they do), the use of this technology does not appear to have significantly affected their organizational structures.

Although these companies do provide a great deal more flexibility to their employees than the typical firm, their policies are also strongly motivated by economic concerns rather than quality of work life. There is very strong competition for a set of unique skills and employee demands are considered because of the necessity to retain their skills. The employees to whom the opportunity is given represent a small proportion of total workers, and fit the "privilege" profile described earlier.

Moreover, there was little evidence in the groups studied of responsibility for the basic operations of the firm (sales, manufacturing, ordering, etc.) There was no evidence, for instance, that those in accounting, manufacturing, or sales were permitted the same flexible work options as those in development. Although development work is probably the most significant operation relative to company success or failure in the industry, it is not tied to the day-to-day operations of the firm and thus can be given more flexibility without much day-to-day impact.

The employees' lifestyles and attitudes toward work reflect their "privileged" status. It is already clear that they have skills for which there is considerable competition. They seem to take advantage of this situation by making demands to fit their own personal "lifestyle" preferences. The California lifestyle in general encourages individualism and unusual recreational pursuits.

The individuals who work at home, like many of their peers, have moved frequently from one company to the other within the industry and the geographical area. They exhibit strong loyalty to their product and their work group rather than to their company, and value highly individual autonomy over organizational commitment. Companies tolerate these attitudes as well as demands to work at home because of their need for the special skills the employees have to offer.

In summary, there was little evidence that companies in the future will encourage work at home for the same reason as the high-technology companies today. The situation represents exceptional employees in exceptional organizations, and no generalizations can be made about the role of technology in changing organizational structures and increasing organizational flexibility.

Predictions for the Future

Based on the findings of this study, work at home appears to be an example of two potential trends toward altering the relationship between employees and organizations, based on economic and social phenomena and facilitated by technology.

In one direction, it is possible that organizations will become increasingly operational in terms of acquiring skills. They will operate in such a way as to purchase specialized skills rather than invest in human resources. There will be continued increases in contract work and individuals will become increasingly autonomous in their relationships to their employers. Those with skills that are in demand will benefit economically; those with skills in oversupply or no particular skills will suffer. Those who see themselves as "new entrepreneurs" will take to the arrangement, but the potential for exploitation is also very real.

In the other direction, work at home may be viewed as one of many innovative mechanisms companies employ precisely as part of their commitment to investing in human resources and improving the quality of work life. Work at home is permitted because management shows trust and commitment in its employees; other flexible work options encourage employees to maintain high productivity for the organization while accommodating their needs outside of the work domain. It is precisely this kind of commitment to quality of work life that is so important to encouraging innovation and maintaining a competitive edge.

We suspect that in the next few years work at home will grow in both directions. We see advantages for both employees and organizations in both. We sincerely hope, however, that organizations implement work at home and other flexible work options as mechanisms for accommodating individual needs rather than exploiting them.

II IMPLEMENTING TECHNOLOGY IN THE OFFICE

9 Understanding the Implementation of Office Technology

Tora K. Bikson
The Rand Corporation

ABSTRACT

What leads to the successful introduction of new office technology in an organization? This question was addressed in a survey of 55 work groups using advanced office technology in 26 organizations. Success includes the extent to which the technology is used, the users' satisfaction with it and with the jobs they perform using it, and improvements in organizational performance. Features of the organization itself, features of the technology, and the process by which the technology is introduced into the organization all play a role. For example, work groups in which computerized jobs retain variety, in which workers have exclusive use of a workstation, have electronic mail, and functional software, and in which the implementation used a balanced social and technical approach and encouraged worker participation in the introduction process all had more successful implementations. The chapter concludes with observations about areas where technology, implementation process, and research all need improvement.

The 1980s have witnessed the widespread introduction of computer-based tools into white-collar work, and the forecast is that this rain of new technologies will continue for the remainder of the century. For this reason, we undertook a field research project aimed at exploring how computerized procedures are incorporated in information intensive environments. The study was supported by the National Science Foundation (see Bikson, Gutek, & Mankin, 1985) as part of its general interest in technology transfer—how are innovations successfully transferred from contexts of development to contexts of application?

We approach the understanding of innovative office technology from this standpoint, treating implementation as a proper instance of "the translation of any tool or technique, process or method of doing, from knowledge to practice" (Tornatzky et al., 1983; cf. Bikson, Quint, & Johnson, 1984). Given that organizations will be deluged with electronic information tools in the years to come, it would be very helpful to find out whether there is something generalizable to be learned about the transfer and utilization of these new work-transforming technologies.

CONCEPTUAL FRAMEWORK

For this purpose, we relied on a conceptual framework that has proved relatively robust in recent research on the implementation of organizational innovations across varied domains (see Berman & McLaughlin, 1978; Bikson, 1980; Eveland & Rogers, 1980; Tornatzky, Fergus, Avellar, Fairweather, & Fleischer, 1980; Yin, Quick, Bateman, & Marks, 1978; and the detailed review in Bikson, Gutek, & Mankin, 1981). A review of that literature led us to three classes of hypothesized influences on the successful implementation of office technology.

- Characteristics of the organizational setting—of offices, in this instance—were expected to influence the results. Here we defined *offices* as organized behavior settings comprising at least four persons, at two or more occupational levels, jointly involved with some information-based process or product. This definition is congruent with conceptions of work groups or primary work subsystems employed in traditional organizational research.
- Features of the information technology were likewise expected to affect implementation success. For research purposes we specified the technology of interest as multifunction interactive computer systems deployed to support white-collar work in offices as previously defined above. We did not target specific hardware configurations, system architectures or software applications.
- Third, we explored implementation strategies—the sequence of activities undertaken in the effort to embed new technologies in extant organizational settings. Ranging from initial planning and decision-making to post-installation modification and training, properties of the implementation process itself were expected to exert the strongest influence.

These potential sources of effect were regarded as closely interrelated and situation dependent. Although they can be separately considered for heuristic purposes, they are likely to be reciprocally influential in fact.

The last element required for the conceptual framework—a viable idea of implementation success—was the most problematic. No consensual criteria were available either in research studies of organizational innovation or in more specific literature on office performance. The difficulty was compounded by the fact that we needed criteria applicable across quite varied office settings. We therefore adopted a broad approach to the construct of successful implementation, representing it in a number of ways.

We treated system use, first of all, as at least a necessary condition for success; thus we examined, for example, the ratio of actual to potential users as well as the proportion of potential users who expected to be interacting with the computer system in the near future. We also took user satisfaction (e.g., perceived "friendliness" of the computer system, overall satisfaction with the new technology, and general job satisfaction) as a face-valid indicator of success. Finally we considered a set of performance effects, including user assessments of productivity as well as manager-rated productivity improvements and value-added gains. Together these variables provided a multifaceted representation of implementation success.

PARTICIPANTS AND RESEARCH PROCEDURES

Private sector organizations that had introduced multifunction interactive computer systems as work tools in white-collar departments at least 6 months prior to the start of the research were sought for participation. Table 9.1 describes the sample we recruited (see also Bikson & Gutek, 1983a). It should be noted that in most organizations, advanced office technology is not evenly distributed; rather, some departments have come to think of it as a way of life, whereas others have only just heard of it. The departments participating in this study are early adopters of information technology; thus they do not necessarily represent all white-collar work groups. However, because the sample is otherwise extremely varied (e.g., in mission, hierarchical status, industrial sector), we believe the implementation lessons it presents may be generally instructive.

TABLE 9.1
Research Sample

- 26 Organizations
  - 50% manufacturing
  - 50% service
- 55 Work Groups
  - range: 4–40 employees per group
  - mean size: 10 employees per group
- 530 Employees

Because we sought comprehensive information in a number of domains, we used multiple instruments and data sources. Information-gathering methods included standardized surveys to all members of participating work groups and structured interviews with the department heads; in addition, we collected archival information about the computer systems in use and related budgeting and staffing patterns. A year later we made follow-up research visits to a balanced subsample of about half the participating sites (see Bikson, Gutek, & Flankin, 1985, for a complete description of the procedures). The discussion that follows is based on the first round of research visits to the full sample, although in many instances what we learned during the follow-up stage of the study assisted us in interpreting the findings.

The definition of *office* just given requires that work-group members be organized in relation to information-related processes or products. After data collection we first attempted an a priori categorization of work groups according to their organizational missions. We found that, although the groups varied broadly in respect to task domains, they seemed to be classifiable into four work-group types based on the information-related function they performed. Table 9.2 shows the distribution of participating work groups among the four categories (these are roughly comparable with Chamat's, this volume, three groups, except we have split professional offices into two categories).

The upper management and administration groups in this classification are typically fairly high in organizational hierarchies and have decision-making, planning, and oversight responsibilities. They include, for example, corporate strategic planning offices and fiscal comptrollers' offices. We distinguished two types of groups that carry out professional functions. The "text-oriented" professional offices were so designated because their major products tend to be text-bearing documents (either print or electronic). Examples are legal offices and public relations offices. In contrast, the "technical" professionals tend to have products that take the form of specifications, designs, formulas, and the like. In this category, for instance, are electronic design departments and internal Research and Development (R&D) departments. The fourth group comprises offices that provide support services and tend to be located nearer the bottom of the organizational hierarchy. Reservations and bookings offices, inventory control, and accounting offices are examples in this category.

TABLE 9.2
Work-Group Types

- 24% management, administration
- 29% professional (text orientation)
- 20% professional (technical orientation)
- 27% secretarial, clerical, technical support

We underscore the importance of the work-group level of analysis for attempting to understand the transition to computer-based information tools. First, organizations are actually structured this way, allocating functions to subunits; work-group membership (as the previous examples suggest) has at least as much influence as occupational level on tasks performed (Bikson & Gutek, 1983a). Second, sound decisions about what hardware and software tools are needed cannot be made apart from understanding the type of work unit in which they will be used (Bair, this volume). Finally it is at the work-group level that implementation processes are typically carried out—it provides an appropriate way of chunking the transition.

REPRESENTING IMPLEMENTATION SUCCESS

The theme of good tools for information work broadly expresses the implementation goal for organizations taking part in this study, although they differed in many specific respects. For example, interview responses established that in half the groups new computer technology was viewed primarily as a tool for upper management and professionals; in another third, its usefulness to managers, professionals, and support staff equally was emphasized; and in 17% of groups it was regarded chiefly as a support staff tool. Given the diversity of technology aims and work-group missions represented in the research, we explored many general indicators of implementation success along the dimensions of use, satisfaction, and performance.

Members of participating work groups were all surveyed, whether or not they were computer users. We took as a prima facie index of implementation success the proportion of employees in each who interacted directly with a computer system in some way during their work. Those not currently using a computer were asked whether they expected to within the next year or so. Table 9.3 presents the proportion of current users, future users, and committed nonusers within occupational level. Although the dependent measure used for analysis purposes is proportion of computer users within a work group,

TABLE 9.3
Employees Who Use or Will Use Computers at Work

	Currently use a computer	*Expect to use a computer*	*Do not expect to use a computer*
Executive	36%	46%	18%
Managerial	71%	26%	3%
Professional	79%	18%	3%
Technical	81%	18%	1%
Secretarial	64%	23%	13%
Clerical	73%	21%	6%

their breakdown within job category is instructive. It appears there is not much support for the view that managers will not keyboard, and the majority of executives in this sample expect to be using a computer in the near future. On the other hand, it is executives and secretaries who are disproportionately represented among the committed nonusers.

In addition to percent of users in a group, we constructed two more use-related variables. System acceptance is an ordinal variable with a value of 0 for subjects who say they will never use a computer and a value of 3 for current users; those who expected to learn within the next few months or years received intermediate values. System incorporation, another ordinal variable, is based on users' answers to a question about the extent to which computer-based tasks had become a regular part of their work repertoires; their responses ranged from 1 to 3, while nonusers were assigned a value of 0. Work-group means served as dependent measures in subsequent analyses involving these variables.

In addition to employee use measures, we also explored employee satisfaction measures. Both popular media and trade journals had advanced the view that technology interfaces are so ungraceful and cumbersome (if not frightening) as to seriously impede casual users (Bikson & Eveland, 1985). To explore potential negative effects of interactive tools on users' task experiences, we asked work-group members to rate their level of satisfaction with the "friendliness" of their computer system in particular and with the new office technology in general. Table 9.4 presents their responses. (Job satisfaction data, not tabled here, were obtained by summing responses to two standardized survey items from the University of Michigan's organizational assessment package. Outcomes were similar but more positively skewed and exhibited less variance.)

Examining the data in Table 9.4 suggests that employees in our sample, on the whole, find the electronic information tools introduced into their work settings relatively satisfactory. Although we observed little in the way of differences between work-group types on these and other satisfaction meas-

TABLE 9.4
Satisfaction Reports

	Percent of Respondents			
	Very Satisfied	*Somewhat Satisfied*	*Somewhat Dissatisfied*	*Very Dissatisfied*
Computer system friendliness	40%	47%	11%	2%
New office technology	38%	49%	7%	6%

ures we explored (Bikson & Gutek, 1983a), there are two exceptions. First, technical professionals—those with greatest expertise—are least satisfied with the friendliness of their computer systems. Second, text-oriented professionals are happier with the new office technology than are other types of groups.

Work-group performance effects constituted the third sort of success indicator we investigated. In Table 9.5 a set of productivity-related outcomes are summarized, based on data obtained from heads of participating work groups. When productivity is defined in terms of increased output by the work group, substantial improvements are cited. On the other hand, in terms of decreased labor costs, over half the groups report no change.

Examining the staffing impacts of computerization by job level helps explain why. It is only at the secretarial/clerical level that sizeable staff reductions have occurred, but these positions are least costly to retain. In contrast, managerial positions have remained relatively stable and professional staffs have grown. In general, outcome measures suggest some changes in work performance but do not provide evidence of dramatic overall improvements (see also Bikson, Gutek, & Mankin, 1985).

Effects of computer-based tools on office costs and outputs are displayed here because they represent the most frequently cited objectives for introducing the new technology. To create a dependent variable representing successful implementation across work groups, we combined cost and output reports from managers. The resulting three-level productivity improvement indicator receives a value of 1 for groups who report no positive change relative to their objective (either cost reduction or output increase) and a value of 3 to those who report substantial positive change; intermediate gains receive the

TABLE 9.5

Reported Work Group Performance Effects

Output	***Percent of Groups***		
No increase	13%		
Some increase	47%		
Substantial increase	40%		
Labor Cost:	***Percent of Groups***		
No decrease	51%		
Some decrease	34%		
Substantial decrease	15%		
Staffing Impacts By Job Level:	***Percent of Groups***		
	Less	***No Change***	***More***
Secretarial/clerical staff	36%	60%	4%
Professional staff	7%	73%	20%
Managerial staff	6%	81%	13%

intermediate value. Managerial interviews also probed potential value-added effects not captured by productivity indicators (such as improved accuracy in planning or forecasting, more timely completion of information intensive projects, and the like). To explore such outcomes, we created a three-level dependent variable representing managers' assessments of value-added benefits achieved. Finally, from users' self-reports of the impact of the computer on the quantity, speed, type, and quality of their work plus the quality of their work experience we constructed a summary index of individual performance effects; a continuous variable, its values were generated by factor scoring procedures. (See Bikson, Gutek, & Mankin, 1985, for more information about dependent measure construction.)

EXPLORING INFLUENCES ON IMPLEMENTATION OUTCOMES

Within the three broad outcome categories—use, satisfaction, and performance—we explored a number of ways of representing implementation success. As the preceding section describes, we devised three variables within each category for a total of nine dependent outcome measures. Adhering to the proposed conceptual framework, we next explored each of the hypothesized sources of influence (characteristics of the organizational context, the computer system, and the implementation process itself) to see how they might be related to implementation success.

For this purpose we carried out a number of univariate analyses, looking at relationships between outcome measures and each of the many potential explanatory variables suggested by the conceptual framework. Our aim was to find subsets of variables in each hypothesized component of the framework that (a) have a conceptual basis in previous innovation research; (b) are not confounded with group type; and (c) have predictive relationships to multiple measures of implementation success. Variables that met these criteria within each hypothetical component of the implementation framework were combined and treated as predictors in multiple regression analyses, with each implementation success measure serving as a dependent variable.

All these analyses were carried out at the work group level (number of cases = 55), with individual survey responses aggregated and linked to interview and archival data. Results are reported as significant if the probability value associated with the statistical test employed is .05 or less; obtained probabilities of about .10 or less are termed marginally significant. Findings from these analyses are summarized later (for a more complete account, see Bikson, Gutek, & Mankin, 1985).

All three classes of theoretical influence evidenced some significant relationships to the success measures we studied. For brevity the following discussion focuses on the predictors within each class of influence that had greatest statistical effect across the dependent measures. Findings should be interpreted as indicating systematic associations that are not necessarily causal in nature.

ORGANIZATIONAL CONTEXTS AND IMPLEMENTATION SUCCESS

In order to learn what kinds of activities characterize information-intensive settings and whether they are differentiated by group type, we asked survey respondents whether they performed each of 18 different tasks as a regular part of their work. The task list represents activities that occur frequently in white-collar work and that can be done with or without a computer. Using a factor analysis (see Bikson & Gutek, 1983a, for a more complete description of this procedure) we identified four sets of co-occurring tasks that account for about 60% of the variation in activities performed. The groupings in Table 9.6 show obtained task sets (i.e., tasks that are, on a statistical basis, likely to co-occur). We then assigned scores to individual employees representing the extent of their involvement with the activities in each set.

Task sets strongly differentiated individuals employed in the four different types of work groups. Analyses of variance treating group type as the

TABLE 9.6
Information Handling Activities Co-occurring Task Sets

Factor I	*Factor II*	*Factor III*	*Factor IV*
Maintain files	Edit and rewrite	Fiscal operations	Programming
Process records	Proofread and correct	Statistical computation	Maintain a database
Fill in forms	Write original material	Distribute information	Statistical computation
Handle messages		Maintain a database	NO communication
Keep activity logs		Develop forms	
Maintain inventory		Communication	
Keyboard text or data		Administrative support	

independent variable and factor scores as dependent measures (Bikson & Gutek, 1983a) yielded the following results. Activities in the first group, primarily involving the management of information, are distinctive of individuals employed in support groups. In contrast, individuals in text-oriented professional groups more often perform the writing, editing, and proofing activities in Group II, whereas technical professionals are distinguished by their engagement with the more computational, statistical, and numeric tasks in Group IV. Finally, higher level management and administration employees are differentiated by the information-handling activities in Group III. Interestingly, verbal communication was strongly and positively associated with management tasks but strongly and negatively associated with technical tasks.

We regard these findings as corroboration of our original classification of offices into group types based on organizational function. Although we were not surprised to learn that different types of groups tend to do different sets of activities, we were surprised to learn how many activities are performed in common by white-collar employees, regardless of the type of office. Table 9.7 shows the most widespread information-handling tasks along with the percent of survey respondents who do each in the course of their regular work. Other chapters in this volume (Bair, Bjørn-Anderson, Chamat) suggest some of the reasons for this finding—support staff are performing more professional functions and professional people are handling more of their own support needs. Almost everyone has to communicate either verbally or by text, and most people have some sort of information files to maintain. Looking only at the sum of tasks performed, we found that management and administration employees engage in the greatest number of different activities and support staff, the least. It would seem that work groups differ more in the variety than in the specific type of information-handling function performed.

Moreover, all the work groups appeared to be engaged in tasks that could be assisted by interactive information tools. We then explored a variety of

TABLE 9.7
Information Handling Activities

Most Common Tasks	*Percent who do each*
Communicate verbally	96%
Write original material	66%
Proofread and correct	63%
Edit and rewrite	57%
Maintain files	57%
Handle messages	49%
Fill in forms	48%
Distribute information	47%

organizational variables as potential predictors of successful technological innovation, including context measures (e.g., size, group type, industrial sector) along with organizational structure and job design characteristics (e.g., centralization, autonomy, comfort, advancement opportunity, and the like) for which standardized assessments have been developed by Quinn, Cammann, and their colleagues (see Quinn & Shepard, 1974) at the University of Michigan. A full description of these variables and a more detailed presentation of analyses is available in Bikson, Gutek, & Mankin (1985).

The most promising of the organizational variables studied—11 in all—were then examined in a series of regression analyses treating each of the nine implementation success indicators in turn as dependent measures. Obtained F values ranged from 0.73 ($r^2 = .16$, n.s.) to 8.33 ($r^2 = .68$, $p < .0001$), yielding three significant and three marginally significant findings. Organizational predictors showed weak associations with performance outcomes and somewhat stronger relationships to use measures. They evidenced generally strong relationships to satisfaction variables, being highly significant predictors of satisfaction with the technology ($r^2 = .44$, $p < .001$) and with the job ($r^2 = .68$, $p < .0001$).

Having looked at the regression analyses separately, we next looked at them collectively for patterns among the predictor variables. That is, we wanted to find out which (if any) of these organizational characteristics made systematic contributions to dependent measures across outcome categories. Among them, it is job design characteristics that seem to be most clearly implicated in implementation success. Specifically, variety and challenge in work along with adequacy of resources for task performance emerge as relatively important positive predictors for at least some measures of computer use, user satisfaction, and performance improvements. It is not surprising to find resource adequacy associated with implementation success. Findings for variety and challenge are more noteworthy: They would seem to suggest that the "rationalization" of work and deskilling of jobs can act against the best interests of both employees and employers.

In general, however, the organizational variables studied here do not account for a great deal of variation in implementation success in comparison to the other classes of influence studied. We do not think this means that user context characteristics are unimportant but rather that better dependent measures are needed. We explored most of the standardized survey scales used to represent characteristics of organizations. Although we invariably found high scale reliability (alpha coefficients in the .80s), we are uncertain of scale validity for office work. Most measures of organizational characteristics in the research literature were developed and validated with blue-collar samples and their extension to information workers is problematic. (For instance, an item on the autonomy scale is "I can take breaks when I want," which we suspect has a quite different meaning for office employees than for employees in manufacturing plants.) Traditional archival measures of organizational variables (e.g., frequency of tardiness, absence from work without an illness excuse, or grievances as indices of job satisfaction) also are difficult to obtain in white-collar settings, especially for higher level employees. Further, problems of response bias (generalized positive or negative response set) should be mentioned with self-report variables, although we found our respondents to be fairly discriminating. Finally, it is possible that some of the most important context characteristics for promoting innovation are the most difficult to operationalize and lack standardized measures. For example, encouragement of initiative, experimentation, independency and other organizational culture values (see Bikson, Stasz, & Mankin, 1985 for an example) may well have strong impacts; but these are not captured in the existing store of organizational variables for use in empirical research.

COMPUTER SYSTEMS AND IMPLEMENTATION SUCCESS

Characteristics of the computer systems introduced into the office settings studied also were expected to have an influence on implementation success. New installations characterized nearly a third of the participating groups who acquired them in 1982; this equipment had been in use for less than a year prior to data collection. In contrast, some groups had used interactive computers for several years, as Table 9.8 shows. The equipment represented a number of manufacturers. The vendors most commonly found include Apple, Digital, Hewlett-Packard, IBM, Wang, and Xerox, with equipment from many other makers less frequently observed. We were surprised at the small proportion of single-vendor sites—most groups have equipment from two or more (see Table 9.9).

The distribution of processing power among work groups is likewise diverse. Archival data revealed the following sorts of computing arrangements. The percent of groups in each arrangement in Table 9.10 sums to more than 100 because the alternatives are not mutually exclusive. Most of the work groups have multiple sources of computing support, reflecting the

TABLE 9.8
When Work Groups First Acquired Advanced Technology

When First Acquired	*% of Groups*
1977–79	15%
1980	21%
1981	33%
1982	31%

TABLE 9.9
Number of Different Vendors Represented in Work Groups

No. of Vendors	*% of Groups*
1	24%
2	26%
3	19%
4	22%
5	8%
6	2%

entrance of personal computers into settings already served by some kind of central processing unit.

Our exploration of technology characteristics as influences on successful innovation followed the analytic strategy we pursued in studying organizational characteristics (see prior section). But because there is not a long tradition of behavioral research involving computer system characteristics, we had to develop independent variables for this component of the conceptual framework as a part of the research project. We approached this task from two directions, using both survey data and archival information supplemented by interview responses. In both instances we followed Bair's (1980, this volume) work by starting with detailed feature lists and constructing more generic variables from them. Our aim was to construct variables that met the criteria specified above, as well as two others; they needed to (a) be independent of particular vendors and specific applications, and (b) represent properties of computer systems that have a relatively direct impact on users.

Work-group members, as part of the survey, were asked to rate the satisfactoriness of specific properties of their computer systems. To learn whether these properties could be combined to form more generic variables we again carried out a factor analysis (see Bikson & Gutek, 1983a). The four resulting factors and the features that define them are presented in Table 9.11; together the factors account for more than 60% of the variance in user responses to

TABLE 9.10
Source of Processing Power for Work Groups

Computing Support	*Percent of Groups*
Microcomputers	31%
Mid-size computers	52%
Mainframe computers	50%

TABLE 9.11
User-Generated Dimensions of Information Technology

Factor I

- Text or data alteration capability
- Text or data entry capability
- Organization of stored information
- Information retrieval capability
- Appropriateness for your specific job functions
- Error detection and correction
- Back-up to prevent file loss
- Keyboard layout

Factor II

- Promptness of maintenance
- Quality of maintenance
- Quality of printout
- Quality of the video display
- Back-up to prevent file loss

Factor III

- Quality of the operating manual
- Type of dialog with the computer
- Response time of the computer

Factor IV

- Convenience and comfort of office furniture
- Arrangement of equipment, furniture and space

computer system characteristics. We construe the first factor chiefly in terms of functionality, or goodness of fit between software design and job functions. The second factor, in contrast, emphasizes equipment performance. We interpret the third factor as standing for interaction support—whether users have what they need to take advantage of interactive tools. The user environment is represented by the fourth factor. We regarded these factors as major dimensions of multifunction interactive systems from the user perspective. Accordingly we converted them into predictor variables for further exploration (using a standard factor scoring procedure and averaging the scores within work groups).

Next, we attempted similar procedures with objective (versus perceived) properties of computer systems. Although these properties are "objective" in the sense that they are independent of users' experiences of them, associated variables and measures reflect researcher choices. For example, the number of different vendors represented in a work group's equipment could be summed and treated as a continuous variable (a proxy for tool diversity with the advantages and disadantages that entails). We also wanted to represent system complexity in terms of software functions. However, we found we could not simply count and sum the number of different applications without making very arbitrary assumptions. We decided therefore to treat system complexity as a three-level variable (low, medium, high) rated by researchers on the basis of number and type of functions implemented.

Examination of archival data, in contrast to survey data, did not yield a limited and comprehensive set of computer system dimensions. We therefore examined separately a number of hardware and software features that fulfilled the criteria previously suggested and seemed potentially capable of affecting implementation success, adopting an exploration strategy very much like that employed with organizational characteristics. These investigations and trial analyses with many computer system characteristics variously represented yielded eight archivally based variables for further study: shared workstations, electronic mail, system complexity, customized software, user modifiability, number of different vendors, use of microcomputers, and use of minicomputers.

Survey and archival variables—12 in all—were then treated as predictors in regression analyses; as before, the implementation success indicators served as dependent measures. Obtained values of F ranged from 1.14 (r^2 = .26, n.s.) to 5.26 ($r^2 = .62$, $p < .0001$), producing six significant and one marginally significant results among the nine regression analyses. Again, satisfaction outcomes were easiest to account for, overall; however, in contrast to organizational characteristics, system variables were more strongly related to perceived system friendliness ($r^2 = .56$, $p < .01$) and satisfaction with the technology ($r^2 = .55$, $p < .001$) than to job satisfaction ($r^2 = .41$, $p < .05$). In the use category, system properties were marginally associated with percent of users ($r^2 = .34$, $p = .10$) and strongly associated with incorporation of system use into task processes ($r^2 = .62$, $p < .0001$). With respect to performance, the system characteristics studied accounted for 47% of the variance in user-assessed work impacts ($p < .01$) and 42% of the variance in managers' ratings of value-added benefits ($p < .05$).

Next, we looked for patterns within the set of predictor variables, attempting to see whether any of them was systematically related to implementation success across dependent measures. The strongest single predictor variable in the regression equations is whether or not a workstation is shared; exclusive use makes a highly significant contribution to extent of incorporation of interactive tools into task processes, satisfaction with the new technology, and user-rated productivity improvements. Other important variables derived from archival data are having electronic mail (the system property most strongly associated with proportion of group members that are system users), customized software, and user modifiability. Interestingly, user modifiability makes a marginally negative contribution to perceived system friendliness; however it makes a significant positive contribution to overall satisfaction with the technology and to user-assessed performance benefits. (Table 9.12 summarizes the computer system variables most closely related to implementation success.)

Among survey-based variables, two of the four factorially generated dimensions emerged as consistent positive predictors: functionality (first factor) and interaction (third factor). As might be expected, functionality is strongly related to the extent of incorporation of computer-based tools into day-to-day tasks; it also is a strong predictor of system-related satisfaction measures. The interaction factor makes substantial individual contributions to all three work-group performance outcomes; it also is a significant contributor to system-related satisfaction.

In general, except for having enough workstations to go around, hardware properties seem less important than software properties for predicting implementation success. It is probably inapppropriate to compare the relative importance of perceived and objective system properties (because the former are summary variables but there are twice as many of the latter). In our data, however, the factorially generated functionality dimension seems very much like the user's view of customized software (less than 20% of participating work groups had wholly off-the-shelf software). The interaction dimension also appears like a user-based counterpart of user modifiability (an archival variable). That is, we think there is some convergence between the subjective and objective measures. This is not to imply that there is an unambiguous set of system properties that predict implementation success. As previously noted, there is not a straightforward relationship between the functionality and interaction factors. The same seems to hold true of important objective properties—systems that are customized are not typically easy to modify. These exploratory findings suggest that the implementation process involves implicit or explicit choices between success criteria (e.g.,

TABLE 9.12
Technology Characteristics That Affect Success

User-Generated Properties:
Functionality
Interaction support
Archivally Based Properties:
Unshared workstations
Communication
Customized software
User modifiability

"friendliness" vs. longer term satisfaction with interactive tools). Nonetheless, for the outcome measures used here, computer system variables outweighed the collection of organizational characteristics we studied in trying to account for variation in implementation success.

IMPLEMENTATION PROCESSES AND SUCCESSFUL OUTCOMES

The organizations participating in this research introduced computer systems into their offices because employees at all levels of the hierarchy were engaged in a great many information tasks; they aimed at improving work-group performance in varied ways by providing advanced information tools (see previous). On the basis of previous innovation studies, we believed that characteristics of the processes by which new technologies were introduced into extant work settings would have a great deal to do with the success of these efforts.

Although there is a long history of innovation research, the major concepts it has investigated were not readily translated into variables for this study. Again, we began the exploration by developing measures of implementation process characteristics that are importantly linked to innovation success in existing literature, investigating their relationship to success indicators in a variety of univariate analyses. Implementation process data were drawn primarily from managerial interviews, supplemented by survey responses, researcher ratings, and archival information (see Bikson, Gutek & Mankin, 1985, for a more detailed representation of implementation variable development).

Available innovation research gives considerable importance to stage in the implementation process (Bikson et al., 1981) as an explanatory variable. Initially, any technological innovation in a work setting is thought to disrupt regular work repertoires and stimulate a great deal of change as the organization adapts to new capabilities. Gradually the turbulence is expected to give way to smoother transition processes until finally a new stasis is reached. Although this thesis accords with common sense, it is not clear how to operationalize "stage." Often stage is equated with "age" of an innovation, or length of time since first access to the new tool; however the passage of time can signal delays and errors as readily as progress. Similarly, amount of turbulence (or degree of system change since initial implementation) can be treated as a stage measure; however stasis may be a sign of too early routinization rather than full deployment of system capabilities. And former theories fail to take into account that if the state of the art is rapidly advancing—as with computer-based tools—continuous system evolution rather than stasis in user settings should be expected. Consequently we also included a three-level researcher-rated stage variable, referring to whether computer system use was just beginning, well under way, or fully incorporated into work-group behavior within a setting.

A second key characteristic identified in prior innovation research is user participation in the decision process. Several study instruments included items that tapped this issue. The best was a survey item asking users whether or not they took part in decisions about organizing work in relation to the system; the proportion of positive responses per work group is significantly associated with implementation success across dependent measures. (Managers' judgments about the extent of user participation in decisionmaking are correlated with but systematically higher than users' own judgments.)

Varied other implementation strategy characteristics were also examined, such as focus of efforts (technology focus, group focus, sociotechnical balance), status of the work group with respect to technology diffusion in the organization, voluntary vs. mandatory transition to new tools for employees, and user satisfaction with the level of learning support (see Bikson & Gutek, 1983b, for an extended discussion of training in computer-mediated work settings). In all, a total of 12 implementation variables representing either stage or process characteristics were selected on the basis of preliminary explorations for further study in regression analyses using success indicators in the three outcome categories as dependent measures.

These analyses yielded values of F ranging from 0.98 ($r^2 = .26$, n.s.) to 4.41 ($r^2 = .61$, $p < .001$); the set of implementation process characteristics explained a significant amount of variation in eight of nine dependent measures, leaving only manager-assessed productivity improvements out of account. Once more, all three satisfaction outcomes were readily predicted; the strongest finding was for satisfaction with the new technology ($r^2 = .61$, $p < .001$). Implementation process variables also were associated significantly with use measures; in this outcome domain the strongest regression equation had proportion of users within a work group as the dependent measure ($r^2 = .57$, $p < .001$). Within the area of performance outcomes, implementation process characteristics evidenced significant relationships to users' assessments of performance improvements ($r^2 = .54$, $p < .01$) and to managerial judgments about value-added benefits ($r^2 = .57$, $p < .001$).

As a last step in the exploration, we investigated the regression analyses for systematically influential predictor variables. Consistent with earlier innovation findings, stage of implementation as rated by researchers makes strong individual contributions to system use and work-group performance measures. Age of system (from archival data) shows a very similar but less strong pattern of influence. On the other hand, amount of system change since initial implementation (from managerial interviews) is a distinctly negative predictor. Not surprisingly, the age of a system shows moderate positive correlations with both the stage of its implementation and the degree of change it has undergone in the process (Table 9.13 summarizes implementation process variables most consistently associated with successful outcomes).

TABLE 9.13
Implementation Characteristics That Affect Success

Implementation Stage Implementation Strategy:
Sociotechnical approach
Learning support
Positive change orientation
User participation
Diffusion status

Among implementation process features, a balanced social and technical approach is a strong predictor, contributing particularly to tool acceptance and percent of users as well as to managers' assessments of both productivity and value-added gains. Another important predictor variable is user satisfaction with learning support, associated with incorporation of computer-based tools into task processes and user ratings of performance benefits; it also contributes to satisfaction with the technology. Other implementation strategy elements that make consistent but less strong contributions to outcome measures include the organization's positive orientation toward change (a standardized survey measure from Cammann, Fichman, Jenkins, & Klesh, 1978) and user participation in the decision process. Finally, the work group's diffusion status—its potential role in technology transfer—emerged as a positive predictor. Groups that were official pilot projects were very successful, perhaps because they received a great deal of organizational hand-holding and encouragement. But groups who regarded themselves as leading examples, as innovators in their milieux, also were quite effective. At the other extreme were groups who believed they would have no impact or might even be seen as a negative example. Conversely, the total proportion of work groups in the same site who were also implementing interactive technology yielded a similar but weaker and negative pattern of influence.

Exploring implementation characteristics suggests that in the transition to advanced information technology—like other organizational innovations—the nature of the process can importantly affect the nature of the outcomes. This conclusion merits emphasis in view of the fact that a sizeable proportion of organizations focus their innovation efforts on the technology itself, paying little attention to the change process. For example, most of the organizations taking part in this study did not have a budget earmarked for implementation and could not estimate the cost of meetings, planning procedures, and other hands-on activities associated with the conversion to computer-based tools. Further, although our findings are coherent with prior research in highlighting the influence of implementation process characeristics, they diverge in suggesting that a successful transition to computer technology will be indicated by organizational stasis; in contrast, we suspect the mark of successful implementation with a rapidly advancing technology may be an organization's learning to manage change rather than attempting to minimize it (Bikson & Eveland, 1985; Bikson, Stasz, & Mankin, 1985).

DISCUSSION

In reviewing what we learned from studying how organizations introduced computer-based procedures into white-collar work, we believe some of the nonfindings are as important as the findings. Among the widely touted problems that we did not observe in our sample, first of all, was resistance to computer use. Many sources (see literature review in Bikson et al., 1981) had predicted that employees would not be willing to interact directly with information systems—some targeting sex (men will not use a keyboard), others citing age (older employees cannot make the transition), and still others basing their expectations on job status (managers won't work at machines). As Table 9.3 shows, none of these predictions was borne out by the data we collected. In fact, rather than finding equipment going unused, we found that limited computing resources was by far the more common problem—work groups reported they needed more workstations and more processing power.

Second, we did not observe software "unfriendliness" to be a major problem. In the main, employees are relatively satisfied with the applications they use (see Table 9.4). This is not to say they have access to the best of interfaces, but only that functionality seems to outweigh friendliness in the work groups we studied. Employees appear to be highly motivated to use good tools, and there is some evidence that friendliness (or, ease of use by untrained individuals) may stand in the way of functionality in the long run (see Bikson & Gutek, 1983a). On the other hand, effective use of a complex interactive tool requires better learning support for employees than many organizations are prepared to provide (see Bikson & Gutek, 1983b).

There are, however, a number of issues raised and not resolved in this research. These issues concern how organizations can make multifunction interactive systems into effective work tools, given that employees are willing and able to use them. For instance, we have observed that although organizations are fairly successful at getting applications to support highly specific functions they have not been notably successsful at supporting generic functions. Table 9.4 provides evidence that there are many white-collar activities performed in common by a sizeable proportion of employees in an organization, regardless of their departmental affiliation or occupational status (e.g., communication, file organization, and management). Future research needs to address the question of how to enhance the generic information and communication environments in which more narrow, task-specific applications are embedded.

Next, subsequent research needs to investigate system functionality for upper level employees. Although over three-fourths of the organizations in our sample aimed at providing support for higher management and professionals, they were most successful at supporting lower level employees (clerical, secretarial, and technical). Major efficiency gains are observed at that level (see Table 9.5), but they have little impact on overall costs or value-added improvements. Further, we suppose this issue is related to improved support for generic activities—because it is higher level groups who perform the most varied but general information and communication functions. Although the question of managerial and professional support has not gone unrecognized, the research reported here suggests two reasons why it has not been adequately resolved.

- Developing "the professional workstation" or "the managerial work station" is probably not a fruitful direction. As we have underscored, the functions performed in common by higher level employees are currently most lacking in support. Second, research at the work-group level has made it evident that these employees do not in the main work as individuals—rather they act as members of teams that include other managers or professionals as well as support staff. Facilitation of shared work by teams of employees, many of whom perform information-related functions in common, is likely to prove more productive (cf. Bair, this volume).
- Developing interfaces to systems that can be used without risk of error by almost anyone, with little training or effort, also is not likely to be effective. We have not found evidence that having to learn how to use a system is aversive to higher level employees (even though quality training is difficult to deliver). Moreover, systems that are "idiot-proof" also are probably competency-proof, preventing well-educated employees from manipulating interactive tools in ways that take advantage of their established skills in task domains. Thus questions of how to increase the power rather than the simplicity of user interfaces merit more attention (cf. Bikson & Gutek, 1983b; Bikson, Stasz, & Mankin, 1985).

Further, there is need for research that explores the links between multi-function interactive systems and organizational performance. Our explorations suggest that where outcomes are readily measurable (in increased output units or decreased input units) they do not represent major organizational impacts. Information and communication—critical elements in an organization's products as well as processes—are not readily assumed under the manufacturing productivity paradigm. Further, we have observed that even when we rely on such measures they are the least readily linked in systematic ways to the characteristics of organizational contexts, computer systems and implementation processes reported here. It is not yet clear how interactive tools affect basic missions and task processes in white-collar work groups, and how these in turn can affect organizational performance. Although examples of dramatic technology-related benefits have surfaced in the literature (e.g., Bikson, Stasz, & Mankin, 1985; Hammer, 1984), we are still unable to define the relationship between the capabilities of flexible, powerful computer systems and the performance of organizations in the information economy.

Finally, there remains a substantial amount of work to be done in developing a research paradigm that will enable cumulative contributions to knowledge about advanced technology in information intensive work. The study reported here is exploratory and suggestive. Although we believe the conceptual framework on which it is based has a great deal of utility for research on technology transfer and utilization, methodological efforts are needed to develop and validate key variables and their measures. In particular, multiple measures of the same variables will help in distinguishing patterns of relationship among the phenomena studied from method artifacts such as response bias in self-reports. More importantly, experimental and longitudinal research is needed to identify causal relationships among variables. As the state of the technical art continues to advance, learning what can be generalized from previous implementation experiences will contribute both to innovation research and to innovation processes in organizations.

10 User Needs for Office Systems Solutions

James H. Bair
Hewlett-Packard

ABSTRACT

Office automation originally was intended to be a solution to the interpersonal communication and information-processing needs of the knowledge worker. However, the cultural forces that drive the mass adoption of technology caused the displacement of office systems by personal computers (PCs). The needs of knowledge workers for communication support and information access have been overshadowed by the rapid influx of PCs. The conclusions from studies of 691 representative office workers define the needs for connectivity with co-workers and information at the work-group, corporate, and public level.

By contrast, interviews at 16 companies to understand the impact of PCs show the need for transparent network connections to corporate information resources, computer administrative services, and user support. The isolation of PCs has generated additional needs rather than solutions. The focus of an office solution is now the work-group computer that provides connectivity and resources to PCs while acting as an interface to corporate resources, and can be operated by a department secretary. The work-group computer completes the office automation solution architecture and provides the shared resources and collaborative tools missing with PCs.

OFFICE AUTOMATION HAS BECOME A CULTURAL PHENOMENON

Office automation is more the result of "culture" than the representation of a planned solution to user needs in the office. Understanding the evolution of computer technology as a cultural phenomenon enables the prediction of future market and product direction.

Culture is the common awareness among people that forms a collective understanding or perception of events. The characteristics of people that influence their perception are similar to the characteristics that define the members of the culture. Language has become one of the most important bases of common awareness. There are numerous other behaviors that characterize a culture ranging from mating rituals to popular sports. Recently in human history, technology has differentiated cultures, for example, automobiles and television. Technological "seeds" of culture start as inventions involving small numbers of specialists, and then as they "catch on" become part of the general awareness.

One method to describe a particular awareness in the current culture is *share of mind.* This refers to the proportion of people who have a similar conceptualization of an idea, person, or technology. In the case of technology for example, the concept with the largest share of mind will dominate the marketplace. The technology, however, may differ considerably from the common awareness. It appears that the greater the share of mind, the larger the difference between the technological solution and the popular conceptualization.

We can use a metaphor developed during the study of technological innovations to illustrate the appearance of different shares of mind. Innovations are most simply represented as a wave (cf. Tofler, 1970). The wave begins to build with the specialized awareness of the technology developers (Fig. 10.1). When persons other than the inventors begin to adopt the technology, it represents an "unfreezing" process. As the technology transfers to the

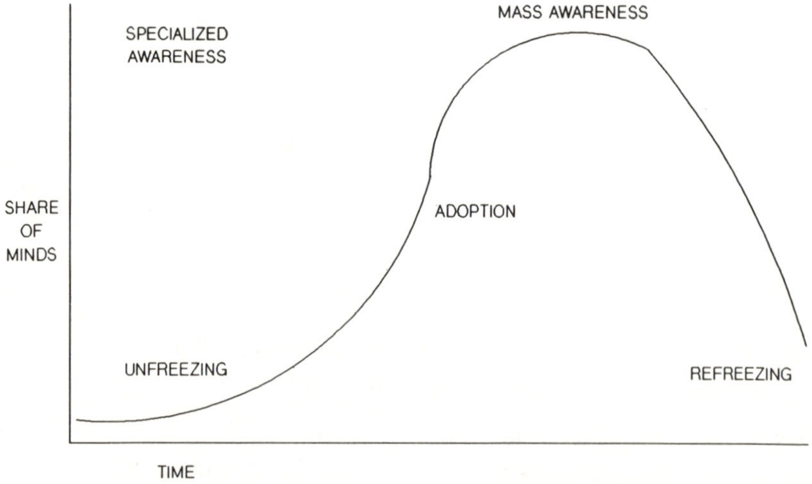

FIGURE 10.1. Technological innovation "wave."

marketplace, it undergoes *reconceptualization* in the users' terms. The outcome at the crest of the wave is a technology, adopted by the marketplace, which is considerably different from the specialized awareness of the original technology developers. All too often, the result is a marketable product, but not a system solution. As the wave subsides, the technology reconceptualization is frozen as production, sales, installed base, and organizational politics all exert massive inertia against change.

Two notable examples in the United States are railroads and the rifle. The inertia of the concept of railroads and the "railroad industry" precluded expansion into other modes of transportation. It is obvious that the specialized awareness of transportation needs was supplanted by the mass awareness of railroading. An additional example of inertia was the persistence of the breach loading rifle after the lever-action Winchester became available. The U.S. Army and many soldiers thought that bullets were better inserted one by one, whereas the Indians, whose mass awareness jumped directly from bows and arrows to Winchesters, had no difficulty outshooting Custer. Custer's last stand may well have been to load his Army rifle.

Fortunately, the model presented here allows for new waves and has the benefit of ever increasing numbers of users. As the awareness of a technology freezes into a myopic conceptualization of a particular product or market, new technologies inevitably appear, initiating a "new wave" that builds toward adoption. In some cases, the new wave may eclipse the previous ones and actually shift awareness from one conceptualization of a technology to another. Certainly, this has been the case with personal computers (PCs), which have involved an ever increasing number of users participating in the exploration of new technology.

CULTURAL FORCES HAVE RESULTED IN OFFICE TECHNOLOGY NOT OFFICE SOLUTIONS

Figure 10.2 presents a greatly simplified view of the evolution of digital information systems. It portrays five waves of innovation based on computing technology: data processing (DP), word processing (WP), office automation (OA), personal computers (PCs), and artificial intelligence (AI). Note that when a wave such as DP falls off, it does not mean that DP ceases to exist. Rather, it shows that the relative number of people whose awareness of digital technology is limited to DP declines and, a larger group emerges with a different awareness, in this case, WP.

During the last 20 years, the awareness of digital technology was DP (remember, "do not bend, fold, spindle or mutilate"), and the initial conceptualization of office automation was WP. However, integrated digital systems for use by knowledge workers were operating in research labs from the

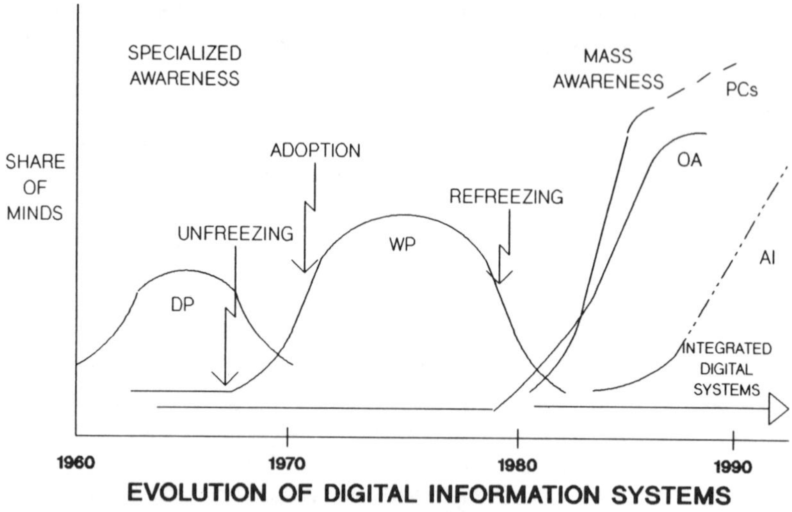

FIGURE 10.2. Innovation waves and the micro wave.

very beginning of this period. It took until circa 1979 for OA to begin to be adopted as it was in the labs during the mid-60s. (Uhlig, Farber, & Bair, 1979).

In fact, at the 1968 National Computer Conference in San Francisco, an office automation system more advanced in many ways than those on the market today was demonstrated (Engelbart, Watson, & Norton, 1973). It presented text, drawings, graphics, and video images on a large screen at speeds over 9.6 Kilobits per second. The system was the Augmented Human Intellect System developed at Stanford Research Institute under the direction of Engelbart, who is now beginning to be recognized as the father of office automation. The system even solved the "problem" of keyboard use by providing a mouse (which has now been "culturally" adopted with help from Apple Computer Co.) and an unusual device that enables one-handed keying of characters as an alternative to the keyboard. The user was able to scan through information as fast as he or she could with paper, keeping his or her eyes on the screen and his or her hands on the mouse and keyset. This system eventually became known as Augment and was used by thousands of specialists throughout the 1970s. Perhaps the most telling aspect of this innovation was the time span from the patenting of the mouse to its adoption. But a small part of the Augment system, this device was catapulted into mainstream culture when it appeared on Apple computers 15 years after it was invented.

Perhaps Augment was too complex to be readily adopted. In fact, the original conceptualization of integrated office systems was too grand to really capture the needed share of mind. With the blessings of the mass media, especially the *New York Times* and *Wall Street Journal,* office systems became a mere shadow of the specialists' dream. Although the idea of OA moved rapidly toward the crest of the wave, spurred on by thousands of consultants, it was soon to be overtaken by more rapidly appearing innovations. Advances in microelectronics resulted in the latest, and largest wave, personal computers.

The functions that were originally espoused as part of OA are now packaged individually as commodity software. For example, WP programs may be purchased from a number of independent software vendors as if WP were a stand-alone function. In fact, it is actually an integral part of preparing information that must be integrated with the production, distribution, management, and control of documentation in general. Originally, OA functions such as electronic mail, conferencing, personal filing, graphics, and others were integrated with editing programs on timesharing computers. Part of the reason for the long delay in the "OA wave" was that professionals were initially seduced into competing with each other for computer power to run these basic office functions. When a timesharing computer had too many users, the response time would become unbearably slow. This was mostly because of cost containment policies that interfered with the provision of adequate computer resources.

Early in the 1980s, a combination of lower cost personal computers and a dazzling "new" function, the spreadsheet, caused PCs to acquire a dominant share of mind. Initially, PCs were not conceptualized as office tools, but they very quickly dominated computer users' awareness and also created an entirely new awareness. The rise of the PC wave with those previously having a low awareness of computing in general resulted in a perception that PCs were office automation. Thus, the mass awareness left computer scientists and pioneer developers confounded by the great disparity between the solution they were building and the concept that actually dominated the market.

Figure 10.2 suggests that AI may become the next wave dominating common awareness, although it may take 10 to 20 years. AI currently faces problems similar to those OA faced in the mid-70s. The specialized awareness is too complex and requires an above average technical background. However, expert systems may become the common embodiment of AI because of concrete applications. As with OA, entrepreneurs are using every means at their disposal to generate a marketplace for AI products, including what might be an overselling of near-term capabilities and relegating the human issues to tokenism. (There is insufficient corporate or government research into the human impacts of OA or AI.)

OFFICE SYSTEMS: THE CURRENT SHARE OF MIND

Figure 10.3 portrays the high-level concepts that can represent both the specialized and mass awareness of office systems today. It illustrates that the concepts that have the largest share of mind, WP and PCs, are only the tip of the iceberg. This perception of WP is left over from the days when the WP wave crested, and it may still influence buying decisions. The role PCs play in the overall awareness may be analogous to the "Model T" Ford, the first mass-produced, mass-consumed automobile.

Below the surface of the water in this illustration are the next concepts that are likely to emerge into mass awareness: mail, messaging, and conferencing. These functions depend on high-speed interconnection of PCs (or personal terminals and a shared host computer) and a critical mass of users. *Critical mass* means that enough users must have access to the capability to make it as worthwhile to use as the telephone. Users must be able to send a message or make an entry in a conference data base, and be confident that the intended recipients will be able to read it as soon as possible. Of course, this means that most intended recipients also must be users of the system.

Data base access from a PC by a non-specialist user is just becoming a reality today, placing it below the surface in Fig. 10.3. Major vendors such as Hewlett-Packard and IBM are now providing connections to data bases on other computers. Multimedia interaction is perhaps further below the surface because of technical requirements to provide the real-time (instant) interac-

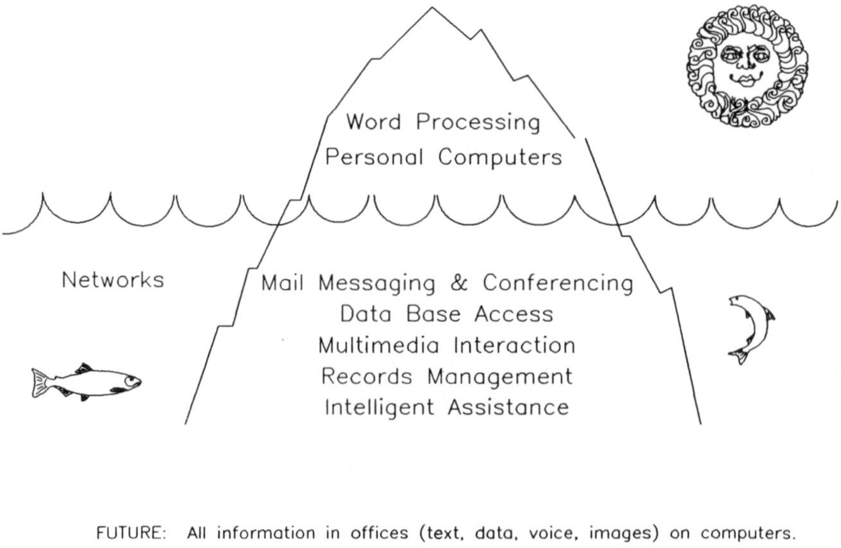

FIGURE 10.3. Office systems current share of mind.

tion between two or more users' workstations. Records management will require the solution of the textual record indexing problem. Ultimately, intelligent assistance may support textual information storage and retrieval, information management, and decision making, but this will require some advances in AI.

Ultimately, all recorded information in offices will be on computers, including text, data, voice, graphics, and images. Perhaps it will take 20 years, but anyone who has held a credit card computer, solar powered and containing thousands of logic and memory components without any additional thickness, will not find it difficult to imagine. In the meantime, the technological push into the market should focus less on technology and more on user needs.

STUDIES OF USER NEEDS AND WANTS

The best way to clarify the disparity between office solutions and the current wave is to present our investigations of user needs and wants.

User Needs versus User Wants

What is needed by office workers may not be wanted. For example, a substitute may be needed for the face-to-face meetings and telephone conversations that create most of the interruptions and consequent disruption in the office. However, this may not be wanted by individuals who enjoy and find it gratifying to interact with colleagues. It has been said that the office is a social institution as much as a place of work. Thus, although the same information may be acquired and transferred through electronic mail and messaging, leading to the same decisions, historically there has been a much lower level of "want" for the communication functions. The following discussion describes office communication behaviors; the payoff for mail and messaging is discussed elsewhere (Bair, 1979a).

When office workers are asked about their wants, they are disposed toward what is currently popular. If surveyed now, they are likely to want PCs with the spreadsheet function and MSDOS. Unfortunately, manufacturers must respond to the popularity of products that are only part of a solution. Consequently, both kinds of studies are discussed here. Needs and wants eventually tend to merge, particularly when there is a pressing need. We are already beginning to see a cultural acceptance of messaging; for example, at Hewlett-Packard there are over 55,000 users.

User Needs Methodology

Beginning in 1979, we decided that user needs required additional verification beyond the sorts of studies currently available (Bair, 1979b). Market research was adequately focusing on wants, but there was considerable uncertainty about committing to product development on the basis of that alone. The new studies focused on needs as the basis of predicting what would become wanted in 2 to 5 years.

The studies conducted by myself and my colleagues at Bell-Northern Research (BNR) were of 691 knowledge workers in seven organizations, ranging from manufacturing to government (Fig. 10.4). In addition to representative industries, the size of the organizations included a full range of staff and revenue. Questionnaires and interviews were administered to a representative sample of knowledge workers (including secretaries and high-level managers) in five of the seven studies. These were complemented by direct observations of workers in their office setting, providing some objective data about office procedures.

An extensive methodological framework was developed from which the questionnaire, interview, and other instruments were designed. Although the complete methodology cannot be revealed because it is proprietary to BNR, a brief discussion is presented here.

The framework was developed based on previous research before the instruments were designed. It represented a multilevel factor list where the top level factors represented 12 categories of investigation, for example, interpersonal communication patterns, time spent in daily activities, and attitudes toward technology. The second level list represented hypotheses about the interrelationships between the factors. For example, it was hypothesized that a self-evaluation of low productivity would correlate with the frequency of interruptions to a respondent's work. The third level list consisted of questions for interviews, questionnaires, or observation. The framework essentially represented a theory of office worker productivity, and it is well documented in internal BNR reports.

	INDUSTRY	SIZE		SAMPLE	
		STAFF	REVENUE	QUESTIONNAIRE	INTERVIEW
CASE 1	R&D	3K	300M	40	-
CASE 2	MANUFACTURING	29K	2000M	-	18
CASE 3	MANUFACTURING	11K	2000M	96	30
CASE 4	FOOD SERVICE	40K	700M	125	13
CASE 5	RETAIL	65K	3000M	150	30
CASE 6	BANKING	90K	20000M	146	16
CASE 7	GOVN'T	7K	500M	134	22
			TOTAL =	691	129

FIGURE 10.4. Studies of knowledge workers in U.S. corporations and government.

Questionnaires. The questionnaires included a variety of response formats in order to compensate for a dependence on self-reports for much of the quantitative data. The questionnaire was divided into several sections that were administered in a prescribed order to minimize interference. In addition, we insured that respondents who had almost no experience with the technology were not asked to predict its impact. Thus, they were not presented with questions about technology, but rather with questions about the desirability of changes in office work. We also asked about any prior source of bias that might confound results, for example, attitudes toward computers and experience with computers. The sections of the questionnaire employed different response formats such as attitude scales, multiple choice, open-ended, and frequency scales. There were over 100 items, but the average completion time was less than an hour.

Interviews. An understanding of the purpose and procedures of office behavior is critical to identifying user needs. We asked about organizational structure, functions, goals, and plans. The interviews were more focused on a particular part of the organization, whereas the questionnaires were intended to be representative of company-wide office work.

Interviews also permitted complex office procedures to be described, which is not practical in a questionnaire. Because our goal was to capture needs, our interview data consisted more of inhibitors to procedural work, rather than detailed accounts of work flow. The purpose of an office behavior was compared to a rating of "value." If a respondent was filling in five forms a day, we asked how valuable this behavior was to the purpose of his or her job or organization.

Observations. In order to maximize response objectivity, we spent a great deal of time in the organization, observing daily working patterns and artifacts. For example, we noted how paper was stored, piled, and routed; how many phones were ringing, what people were doing at their desks, how the office furniture was positioned; traffic patterns to other desks, copiers, service facilities, and superiors; and the overall ambiance and attitude. In particular, we noted the queues at the copier and the backlog at secretary and typist stations. In one of our more intensive cases, we spent 1 day per week for an entire year on site.

Other Methods. The comprehensive methodology included other methods that were not used in every case, and thus are not discussed here. Network analysis was applied to certain departments to understand communication patterns. This method requires respondents to list their top 40 communicants and to rate the frequency of interaction for each medium (meeting, phone, mail). A computer program provides a cross check between communicators and an interpersonal network topology. "Key-product tracking" identified an output from an organizational unit and logged the labor and throughput time required for completion. Other methods were related to our long-range agenda to perform a posttest study after office systems implementation (cf. Bair, 1978).

Respondent Selection. Whereas the organizations were self-selected by their willingness to participate in the joint research effort and to contribute funds, respondents were selected by stratified random sampling. The entire organization was scanned to understand the different functions and job types. In the seventh case, we used a short questionnaire administered to about 2,000 respondents. Thus, we were able to identify internally homogeneous groups from which to sample. For example, a stratification was found between the procedurally oriented functions (purchasing, accounting, and personnel), and the less structured units (marketing, planning, management services). The resultant sample, averaging 99 respondents in each case, was representative of the different types of work and job types (secretaries, managers, professionals, administrators, etc.).

Analyses. Statistical analyses were rigorous within each case, but we did not combine the cases into one data base. There were two reasons for this: First, the instruments were refined in each successive case, and secondly, it was not necessary to combine numerical data in order to draw meaningful conclusions across the cases. The conclusions briefly discussed here were the same for each case and averaged across all cases. For example, the rank ordering of needs was almost identical in each study. Because a primary goal was to identify the predictors of productivity, we used a variety of multivariate analyses to massage the data and develop indices. Correlational matrices combined with conceptual clustering resulted in indices such as personal and organizational productivity. Multiple regression, factor analysis, and analysis of variance identified the predictive relationships. Through index construction, we were able to relate the different response types to homogeneous variables. For example, the cluster of attitude questions about personal productivity was predicted by the amount of time spent in unscheduled meetings (inverse relationship).

Interview data was summarized, reported anecdotally, and compared across respondents. Where interviews and observations contradicted analysis findings, the analysis was considered inappropriate. The average number of interviews was 18 per case, a sufficient number based on the replication of findings within each interview stratification. Interviews and the other methods supplemented the framework in driving the data analysis.

User Needs Findings

We decided that describing what the workers did and where they did it was a prerequisite to any reasonable understanding of their needs. It was found that all office workers spend a majority of their time communicating. This finding is generally corroborated by researchers such as Mintzberg (1973), and Panko (1982) at the University of Hawaii, who compiled the results of 14 studies of several thousand workers.

The percentage of time spent in the general activities, face-to-face meetings, telephone interaction, written communication, and "other," which includes analysis, thinking, and administrative tasks such as copying and filing, are shown in Fig. 10.5

In communication activities the location of senders and receivers is of vital importance to understanding needs. We might suppose, for example, that most written communication is with persons quite remote from the originator, thus placing the emphasis on pieces of solutions such as wide-area networks. Some of the key results are summarized following, and represent averages across those cases where data were obtained (i.e., the N is less than 691 for some of the figures).

- Office workers, on average, spend 65% of their time in interpersonal communication (see Fig. 10.5).
- The majority of face-to-face communication is with the immediate work group (48%). The remainder is divided among the same building (17%), different locations within the company(13%), and external to the company (22%) (see Fig. 10.6).
- The majority of written communication is also within the work group (38%). The remainder is divided among the same building (24%), different locations (22%), and external to the company (14%), and is less than expected (see Fig. 10.7).
- Most paper-based files are stored within the person's office and work group. This is based upon observations of the physical size of storage. When daily working information is included, the greatest need and frequency of access appears to be for the same paper information.

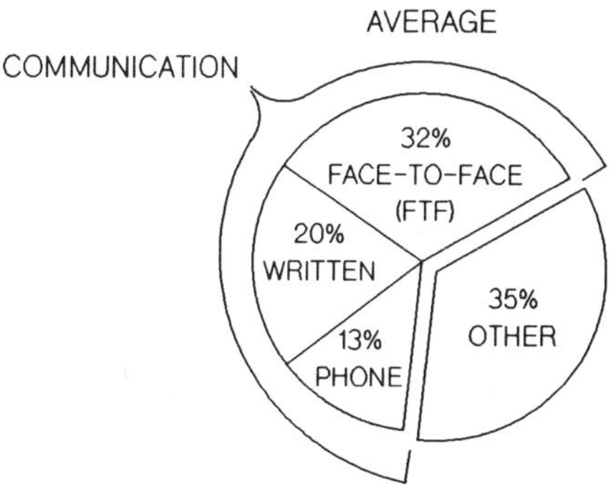

FIGURE 10.5. Knowledge worker activities.

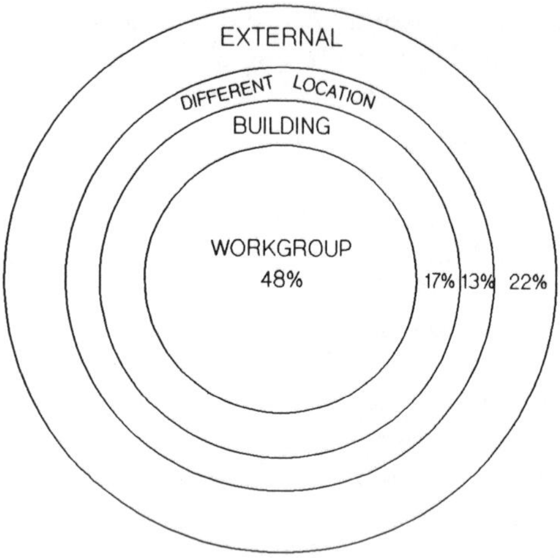

FIGURE 10.6. Knowledge worker locus of communication: face-to-face.

The conclusion from these results is that knowledge worker communication at the work group level is the largest labor investment in the office and therefore should be a major focus of the office solution. The specific composition of the work group varies considerably among organizations depending on formal structure, function, type of work, and size. On average, work groups tend to coincide with the departmental unit, with 10 to 20 professionals and a secretary reporting to one manager. Usually, smaller task groups of five to eight persons exist within the department where there is a slightly greater intensity of interaction.

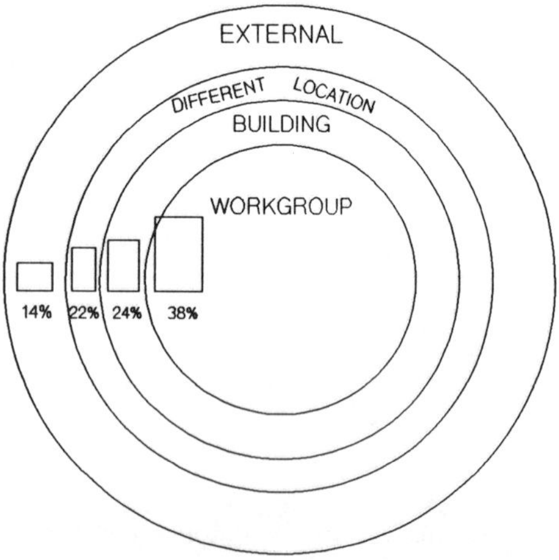

FIGURE 10.7. Knowledge worker locus of communication: written.

Having established what knowledge workers do as a prerequisite to understanding their needs, data analysis showed that communication needs were the most important. Thus, there is a strong correspondence with what we found knowledge workers do and what they need. To summarize, the need in offices is for knowledge worker connectivity.

The data reduction showed the most important specific needs that are presented in order of importance in Table 10.1. The top need, increasing access to remote information, was most often dependent upon communicating with people who could provide the information or knew where it was located. The second highest need, reducing interruptions, resulted from unplanned communications through the telephone and unscheduled meetings. Note that the fourth highest need, "reduce telephone tag," and the highest need, "information access," conflict with the need to reduce interruptions. The other needs listed in Table 10.1 are relatively straightforward.

Table 10.1 shows four columns reflecting where the needs would have to be met: by personal means, assistance at the work-group, divisional, or corporate levels. All the needs are related more to the work group than to other levels of support. A more thorough analysis of the data supported this cursory notion, but it is straightforward if most needs are communication related and most communication occurs within the work group, then most needs would obviously be met by a solution at the work-group level (Culnan & Bair, 1983).

TABLE 10.1

Knowledge Worker Needs ($N = 691$)

(Rank Order)	*Personal*	*Work Group*	*Div.*	*Corp.*
1. Increase remote information access		X	X	X
2. Reduce interruptions		X	X	X
3. Reduce delays in written comm.	X	X		
4. Reduce telephone tag		X	X	X
5. Provide more resources (support)		X	X	
6. Reduce Redundancy (tasks & info.)	X	X	X	X
7. Even out workload		X		
8. Support personal composition calculation, storage & retrieval	X	X		
9. Decrease irrelevant information & contacts		X	X	X
10. Support personal administration	X	X		

User Wants

User needs and wants are considerably different. User needs relate to general problems or impediments to performance based on an analysis of their jobs. A respondent does not have to understand the state of the art in technology to provide valid information about needs. On the other hand, user wants are the specific desires and requests of the individual studied; it is assumed that the respondent is well versed in the available technology. Note that although this may be the case, especially with PC technology, it is highly unlikely that the typical respondent would understand what technology could be available to him or her in the future.

Thus, wants reflect current inadequacies much more than future solutions. They emphasize technology, and also are focused on the work-group level. My colleagues at BNR and I interviewed managerial representatives of 16 companies during late 1983, and found two general classes of "wants": those reflecting "connectivity" requirements and those for service and support of the end-user (the report is proprietary). The 2–3 hour interviews focused primarily on PC usage, based on the assumption that PCs would generate most new user wants. Connectivity refers to the connections or communications capability between users through interconnected workstations, host mainframes, and computers providing specialized services (servers). Ideally, users want a "transparent data home" where the location of data is invisible to the user.

Connectivity Wants. Connectivity wants include the following:

- Information accessibility from remote locations to department records, and to corporate and public data bases with an emphasis on providing portability, such as with PC files, or providing gateways to remote data bases.
- Information sharing, particularly documents, between principals and support staff, among members of the work group, and with others through electronic messages.
- Simultaneous information sharing in support of meetings and phone calls, such as blackboards and flip charts.

Some of the these wants require elaboration. In the first area, accessibility suggests a portable terminal from which the user can dial-in to his or her information systems or a portable PC that permits reading files while "on the road." Ultimately, the user wants to describe the information he needs at the moment, in a "natural" way, and have the connections established for him or her if the information is not local.

Users are becoming increasingly aware of the second area, information sharing. This includes the high level of sharing between a principal and the secretary, which often involves the rough-drafting of memos by the principal on his or her workstation and the formatting, proofing, printing, and so on by the secretary on his or her workstation. Of course, the sharing of calendars, agendas, to-do lists, and so forth requires coordinated access to the information. In fact, calendars require more coordination regarding priorities, value judgments, and plans than is currently possible to support.

File sharing among members of the same work group has become an issue because of the difficulty of accessing floppy disks on other than the user's own PC. It may not be desirable to permit access to PCs through networks because of security issues. But there is a strong want for multiple authors of reports, plans, memos, spreadsheets, and the like to access a common file. This has implications for more than simple access because work-group members want to revise the shared entity. This requires management of versions and some method of controlling concurrency (simultaneous changes to a file). PCs have also complicated electronic mail (messaging) that requires the transfer of special files between PCs, or accessing a "mailbox" on a shared resource.

Perhaps the most challenging want we found was having two or more persons simultaneously view and change information displayed on their PCs regardless of their location. Called *shared screen teleconferencing,* this requires the management of changes by rapidly shifting control from one user to the other. Shared pointer control also was desired, so that each person could move the pointer and have others see it. The idea is similar to a meeting blackboard where all participants have chalk and a pointer. User wants for connectivity are gaining increased attention by researchers investigating collaborative work (Grief, 1985).

Service and Support Wants. We also found a strong demand for services that were traditionally available from the data processing organization, but difficult or burdensome for the typical PC user. They include file maintenance, applications development, transaction management in the interaction of data bases, work group level data bases, and open architectures to permit upgrades, multi-vendor compatibility, and innovations.

- File maintenance includes back-up of floppies in case of damage to the working copy, archival of file information not currently needed but required for permanent retention, and the use of "mass storage", very large devices for archival, backup, and data bases. Users want to end "floppy flipping madness," especially because they often are lax about it, and want to return to being computer users rather than operators.
- Data processing services are viewed similarly, because PC users have lost application development service, transaction management, and powerful data base models. Programming applications may have taken an average of 2 years in the typical data processing center, but at least

if the user needed an inventory control system he or she did not have to write it him or herself. Data bases have traditionally been modified in a tightly controlled manner from "dumb" terminals that ran transactions on the central store. PC users want to download part or all of the data base, modify it, and then upload the revisions. The effect on data integrity was reported to be insufferable, unless some means of managing the uploads can permit an accurate, central store at all times. Users also want data base structures as sophisticated as those found in large, mainframe environments, which is currently not conceivable with the typical PC.

- Users employ the phrase "open architecture" to describe multi-vendor compatibility. Merely because the user has an IBM PC should not mean that he or she can use and/or communicate only with other IBM PCs. It also includes openness to upgrades, so that new versions of software are easily adaptable, and to innovations, so that new systems can be used without technical support, which is often extensive for new systems. Users also want software to be maintained by someone else through remote access to the software, a rather difficult process on PCs.

These represent just a portion of an extensive list of user wants, many of which have been generated by the relatively recent PC explosion.

THE NEXT STEP TOWARD OFFICE SOLUTIONS

Computers were initially adopted at the corporate and divisional level of most organizations. Public level systems such as library retrieval services (e.g., Lockheed's Dialogue), public messaging (e.g., MCI Mail), and information services (e.g., Telidon) are developing but have not captured a significant share of mind. Personal computers represent the largest share of mind, not only now, but in all of the history of computing, with an installed base of over 7 million in 1984.

The work-group level, between corporate–division and personal, is being recognized as the next step toward meeting both user needs and wants. It is driven by user needs for connectivity with co-workers and local information bases, and user wants arising from the shortcomings of PCs as previously stated. Office automation initially solved many of the problems of shared resources and services because the computer was a shared resource operated by an organization providing support services. PCs have essentially resulted in isolated users, islands disconnected from the people and the information required in office work.

It can be concluded that any office system must incorporate all five levels of computing, interconnected with a high bandwidth (at least 1MB per station) local area network, and incorporating gateways to other corporate and public networks.

The levels of an office system are:

1. Personal
2. Work-group (co-workers and department)
3. Division (same building or campus)
4. Corporate (national or international location)
5. Public (services)

The work-group level computer provides the interconnection between the personal level and the others. It is a natural gateway, particularly between the local area net and wider area services. Perhaps more important than many of the needs and wants that can be met at this level is the support provided to a heretofore neglected component of the office, the department. Departments are usually cost centers and the support staff often reports to the department manager. A primary function of the department is record keeping, ranging from correspondence and personnel files to budgets and other project management activities. Placing these records online is a key component that has not been adequately addressed due to software limitations.

Departmental records and the needed and wanted services for work-group performance can be provided on computers that are now affordable and, perhaps even more importantly, operable by the work-group staff. The typical scenario is a work-group computer under the desk of the lead secretary who can operate and maintain the machine due to innovative "user friendly" utilities and advanced hardware that is reliable and unobtrusive.

PCs in a local area network will have transparent access to filing, including archival, high quality printer facilities, and other functions not affordable or easily maintained by each individual user. The ability to co-author documents will be enhanced by having a common file space. Telephone interaction can be enhanced by simultaneous viewing and pointing to the same presentation on several PCs in the work group. Filing may be enhanced by techniques from artificial intelligence applied to the locally shared data bases.

A tremendous base of computer users has been created by the mass adoption of PCs. This large base, more than ever, provides each individual user with a vehicle for connectivity that was not achievable with traditional computer availability. Networked PCs represent the ticket to full citizenship in the information age.

SUMMARY

In summary, three key points have been discussed that warrant the incorporation of the work-group level into OA solutions.

- Driving cultural forces have resulted in a technological wave of PCs that were not originally designed for an OA solution.

- Knowledge workers in the office need interpersonal communication support primarily in their work group, and access to departmental as well as personal, divisional, corporate, and public information.
- Knowledge workers want connectivity with corporate information resources, service, and support.

Meeting user needs requires changes in the process of office work. Improving interpersonal communication means using office technology in place of meetings, phone calls, and mail, where appropriate. Better information access means contacting the people and interrogating the information bases efficiently and on demand. Current approaches to meeting needs focus on installing technology; a "techno-centrism" prevails. The step beyond work-group level technology is "process management": the redesign of office processes focusing on the improvement of ways in which people work. This human focus will be necessary to meet user needs, and is the challenge of office systems research.

11 Social Interaction and Office Communication: Effects on User Evaluation of New Technologies

Jeanette L. Blomberg
Palo Alto Research Center, Xerox Corp.

ABSTRACT

Based on field observations conducted in varied office settings, this chapter explores the interplay between technology and the social organization of the workplace. Differences observed in these office environments suggest that understanding how machine use is integrated into the workplace requires directing attention beyond the human–machine dyad to the network of co-worker interactions that support both machines and office work. Of particular importance is the way in which information about the machine's day-to-day operation is exchanged among end-users, knowledgeable others, and technical support staff. In environments where information about the ongoing behavior of the machine is consistently channeled through an established, recognized individual or group, end-users have available a reliable source for information about both the current status and performance history of the machine or system. These knowledgeable individuals or groups, acting as repositories for information about the operation of the technology, also serve as important links between end-users and technical support staff. The availability of these knowledgeable others, in turn, is related to the organizational relationship between these individuals and end-users, their geographic location relative to the technology and end-users, and the reward and advancement system operative within the organization. Results of this study suggest that realizing the full advantage from new technologies will require more than designing "self-explanatory" machines; it will require conceptualizing human–machine interactions as embedded in ongoing human interactions wherein the "social" significance of the technology is established.

Much has been written in the last few years about the impact of new office technologies on people's work lives. There are those who have speculated that new computer technologies may alienate workers from their jobs or from one another. Others have debated whether these new technologies will concentrate organizational power in the hands of a few or alternatively decentralize access to information and therefore access to power. Still others have noted the potential for the deskilling of jobs and/or the improved career opportunities for the few highly trained. As important as these and other related issues are to the lives of workers and the operation of organizations, it is curious that more attention has not been focused on the ways in which individuals actually accomplish their work in environments highly dependent on new office technologies (Kling & Scacchi, 1982).

Social Structure

Although computer technologies have become integral parts of the workings of many offices and larger organizational units, this seldom has been achieved without some difficulty. Problems with dissatisfied users or less than optimal applications are cited frequently. Those interested in the implementation process typically have conceptualized the difficulties associated with the introduction of new technologies as related to or dependent on characteristics of employees or end-users and characteristics or social structure of the adopting organization, such as the degree of centralization of decision making, employee autonomy, or management style (Mankin, Bikson, & Gutek, 1983; Cheney & Dickson, 1982; Danziger & Dutton, 1977; James & Jones, 1976; Kimberly & Evanisko, 1981). Others have considered the importance of system design, including the degree to which end-users are involved in the design process and in the successful implementation of new technologies (Bjørn-Anderson & Hedberg, 1977; Driscoll, 1978; Kaiser & Srinivasan, 1982; King & Rodreguez, 1981; Lucas, 1975, 1978; von Hippel, 1976; Wynn, 1982). Less frequently considered are the ways in which these technologies become integrated into the day-to-day activities of individual employees. Although characteristics of the organization, end-users, or system design may be related to the embedding mechanisms, they are not an adequate gloss for the actions of individuals that serve to integrate new technologies into the everyday activities of the organization. Social and economic indicators of behavior such as occupational category, role within the organization, salary, or status always are "reanimated or creatively affirmed" (Sapir, 1949, p. 104) in the day-to-day interactions of workers. It is at the level of moment to moment or immediate actions that these social structural categories become observable.

Human–Machine Dyadic Relationship

Approaching the issue of the impact of office technologies on work lives from a different vantage point are those studies that focus on the difficulties individuals experience when using new technologies. These studies are concerned with the cognitive skills needed by end-users and with the characteristics of system and interface design that make for a "user friendly" or "self-explanatory" machine. The focus of these studies is on the human–machine dyadic relationship or on what transpires between human and machine that makes for a "successful" interaction. These studies address questions such as, what are the mnemonic properties of machine commands? do menu driven systems offer advantages to the end-user? and are analogy and metaphor useful concepts to be built into interface design? (Bayman & Mayer, 1984; Card, Moran, & Newell, 1981; Kiger, 1984; Moran, 1981, 1982; Rosenberg, 1982; Young, 1981). Regardless of the particular questions raised, most of these studies do not concern themselves with the fact that these new technologies are used in social environments and that the interactional dynamics of these environments affect the individual user's experience of the technology. In other words, perceptions of "friendliness" or ease of use are not independent of the social interactional environment in which the system is located.

Social Interactional Environment

What seems to be overlooked in many of the studies that deal with the impact of new technologies on work lives, including those that focus on the human–machine dyadic relationship, are the effects of the social interactional environment on the individual's understanding and perception of the technology and the processes by which new technologies become embedded in the ongoing activities of offices and larger organizational units. At the intersection between the social structure of the adopting organization and the individual user's experience of the technology are the everyday social interactions that take place among co-workers (Fig. 11.1).

The approach suggested here cuts between the cognitive and psychological constraints on behavior and the social structural ones. The focus here is on the social situational context in which machine use takes place. In other words, attention is directed toward how individuals accomplish their work in specifiable contexts. From this point of view then, human behavior is understood in terms of "the ways people organize concerted activities in each other's presence" (McDermott & Roth 1978, p. 321). Knowledge and expectations about a new technology are constructed, maintained, and revised in contexts of frequent face-to-face interactions within networks of established

FIGURE 11.1. Relationship among levels of analysis.

co-worker relations. The nature of these interactions is critical to understanding how the technology becomes integrated into the ongoing activities of the organization.

METHODOLOGY

The study of socially situated action requires a methodology that generally is not represented in office automation or organizational research. The level of analysis undertaken requires using a methodology that allows for a detailed description of individual behavior within particular social settings. Highly structured interviews or written questionnaires are inadequate because they only provide restricted, after-the-fact accounts of the behavior in question and frequently are framed in the terms or categories deemed important by the researcher and not the subjects of the study. This is not to say that large surveys focusing on the organizational level do not have their place in office automation research, but only to suggest that they are inadequate for providing a description of the interactional mechanisms by which new technologies become embedded in the workplace. What is required to provide a richness of description that makes interpretable and grounds in actual behavior the relationships found between social structural categories, system design, and user acceptance and satisfaction is a research strategy that includes extensive on-site field observations and a means of documenting the behaviors observed. Also useful as a supplement to field observations are open-ended interviewing or informal discussions with the participants in the study. These interviews and discussions are particularly useful when important interactions take place out of the view of the observer. Participants' post hoc explanations or justifications for the observed behavior can provide additional information.

CASE STUDY EXAMPLES

With this background in mind, two examples are presented to highlight the potential benefits this approach can offer to our undertanding of how new technologies become embedded in the workplace. The first example is a study of user acceptance and satisfaction with a relatively advanced copier/duplicator, the type of copier that has been described as doing everything but making coffee. As copiers become more and more computer-based, they require skill and understanding levels not unlike those required to use other computer-based technologies. The second example is a study of a group of industrial and human factors designers who are using a highly flexible computer system, Trillium (Henderson, 1983), to design the user interfaces for copiers, printers and duplicators. These examples illustrate patterns of social interaction are important to user satisfaction and to the efficient use of a technology in environments where technologies differs widely.

COPIER USER COMMUNITY

The copier study involved on-site observations and open-ended interviewing at 10 office settings. The office sites ranged from a television station, to a department of the state government, to a computer engineering firm. Common to all of the sites was the fact that the copier was casually operated. A casually operated machine is one where individuals use the copier to accomplish certain tasks, but their main job function is not operating a copier. Where copier operation is their main job function, the copier is said to be located in a dedicated environment. This is an important distinction to make and it applies to other office technologies as well. Users will differ from one another to the degree that their job is operating a machine versus using a machine in order to accomplish some other task. In the latter case, using the technology becomes the means to some end and not an end in itself. In addition, as Cuff (1980) and others have noted, casual users tend to use the technology less frequently and therefore typically have less knowledge of machine operation. Consequently, casual users often are focused less on the technology and therefore more appreciative of its apparent invisibility. However, this relationship seems only to hold in cases where the technology is highly reliable or where technical support is readily available to the casual user. Where this is not the case the invisibility of the technology tends to impair the discovery of solutions to machine problems (Blomberg, 1983).

In order to show how the relationship between social structural characteristics of the workplace and user acceptance of and satisfaction with an office technology is related to the interactional behavior of users, this chapter explores the differences in the environment of machine use at two of the copier sites. These two sites were selected because users' overall evaluations of machine performance differed widely, even though identical copiers were in use at both sites. At one site, casual users gave the copier an overall dependability score of 3.6, where 4.0 was excellent, whereas at the other site the dependabiity score was 1.5. The question here is how the social structural features of these two work environments are animated or brought to life in the day-to-day interaction patterns of employees. In particular, the focus is on how information about machine operation and performance is exchanged among co-workers at these two sites.

Social Structural Characteristics

The social structural characteristics that are related to the interactional behavior of co-workers include: (1) the number of users; (2) the number of "autonomous" departments relying on the technology; (3) the relationship between the department responsible for the technology and [a] the end-users and [b] those who support the technology; (4) the range of responsibilities assigned to the support personnel; (5) the rewards received from management for carrying out these technology-related responsibilities; and (6) the physical location of the technology relative to [a] end-users and [b] support personnel (see Table 11.1).

High Dependability Site. At the site where the copier received a high dependability score, the copier was used by approximately 30 individuals from four departments of a computer engineering firm. However, the machine was under the control of a single department, financial administration. There were three support personnel called key operators who reported to the manager of this department. Key operators are employees who are trained to clear jams, clean the machine, and add supplies. The actual responsibilities assigned to or assumed by key operators can vary from one site to the next. At this particular site the key operators were responsible for cleaning the machine and replenishing supplies each morning, assisting casual users with machine problems and calling for technical assistance when required. The key operators sat near the copier in a single work area with only a partition separating them from the door to the copier room. The key operators at the high dependability site were supported by management for performing their duties and perceived their key operator role as part of their job and not an added burden. As one key operator put it, "I don't mind it [being a key operator] at all. Kinda always been one since I came to the

TABLE 11.1
Structural Characteristics of Two User Environments

	High Dependability Site	*Low Dependability Site*
Number of users	30	150
Number of departments relying on technology	4	8
Relationship between department responsible for the technology and		
a. end-users	25% from dept.	$>$ 2% from dept.
b. support personnel	all from dept.	all outside dept.
Responsibilities assigned to the support personnel	complete range	limited
Support from management for responsibilities	strong	weak
Physical location of the technology relative to:		
a. end-users	50% nearby[a]	15% nearby
b. support personnel	all nearby	25% nearby

[a] Those who are nearby are on the same floor as the machine.

company, with other machines. So, it's kinda like just, you know, part of my job."

Low Dependability Site. In contrast, at the site where the copier received a low dependability score, the copier was used by approximately 150 employees from eight divisions of a large department of the state government. Although the copier was located on the fourth floor of a five-story building, employees from throughout the building used the machine. The copier was under the control of the Business Services Office, although the key operators were selected from each of the eight divisions and did not report to this office. A key operator's only responsibility was to assist casual users from his or her division. Cleaning the machine and adding ink were the responsibilities of the Business Services Office. None of the key operators was responsible for the overall operation of the copier. Furthermore, division supervisors placed little importance on the key operator role, generally feeling that the Business Services Office was responsible for the copier. However, because the Business Services Office was located in the basement, four floors away from the copier and was responsible only for cleaning the machine and adding ink, the involvement of this office in the day-to-day operation of the machine was extremely limited. Consequently, because each key operator only received reports of trouble from a small percentage of the total number of users, there was no person or group in a position to monitor the day-to-day operation of the machine. One of the two key operators who sat on the same floor as the machine described her involvement with the copier and with other casual users by stating, "I don't think very many people know I'm a key operator. . . . I don't get that many calls. If I'm in there [the copier room], I'll tell them how to do it [clear a jam]." Even though she was nearer the copier than any of the other key operators she received few requests for help. Her involvement with the copier and with end-users was not substantially different from that of the other casual users. If she was around when there was a problem, she would lend her assistance. Otherwise users were left more or less on their own.

Interactional Patterns

The interactional patterns that were associated with these differences in the social structural environment of machine use are now considered. What stood out above all else were the differences in the ways in which information about the day-to-day operation of the machine was exchanged among casual users, key operators, and technicians. These differences can be illustrated by describing two typical scenarios that took place at each of these sites. These examples highlight differences in the interactional patterns that affected the integration of this particular office technology into employees' day-to-day activities. The scenarios were constructed in part from field observations and in part from personal interviews.

Exploiting the Full Potential of the Technology. At the high dependability site a casual user experiences a problem making two-sided copies from a single-sided original. This is not a trivial task on the machine in question. After a failed attempt where the available documentation is superficially referenced, the casual user walks a short distance to where the key operators are located and asks one of them for assistance. One of the key operators goes to the aid of the casual user, providing the instruction and guidance needed to complete the task. With relatively little effort the casual user learns how to use this particular machine feature. At this site the key operator serves as a source of information about the operation of the machine and is willing to do so because she perceives the task of helping other users as part of her job.

In contrast, a casual user at the low dependability site when asked by an observer, "Why don't you make two-sided copies of that report?" responds, "Oh, I never learned how. I only know how on the other machine down the hall. But it's down today." This casual user has decided it is too much trouble to learn how to make two-sided copies on this machine. The effort involved in seeking key operator assistance is greater for this casual user than for the one at the high dependability site because the key operator selected from her division is located two floors away from the copier. It also is possible that the key operator might not appreciate interrupting her work activities to help with a machine problem because management places little importance on this activity. As one key operator stated, "Now they [management] have more important things in their mind for their employees to do." The result is that at this site some machine features are not exploited by users. Users must be willing to invest a good deal of time and to experience a certain amount of frustration if they are to exploit the full potential of the technology.

These scenarios point to the fact that the ease of use of a technology cannot be isolated from the social environment in which it is located. A technology that on the surface appears difficult to use may be perceived favorably by employees if there exists the support needed to "translate" the technology into terms understandable to the user. This is not simply an issue of developing better documentation for new technologies. The everyday contingencies and uses for a technology cannot be fully anticipated by system designers or technical writers. "Being there" often is necessary in order to make the technology understandable to the user. In environments where this support exists one can expect a much higher degree of user acceptance and satisfaction and a more efficient use of the technology.

A Resource for Machine Performance Information. Another scenario at the high dependability site proceeds as follows. A casual user experiences a jam in the main processor unit of the copier. Casual users at this site are unable to clear jams in the main processor unit because the machine is kept locked. The casual user therefore seeks the aid of one of the nearby key operators. She unlocks the copier and clears the jam. In the process of completing a rather lengthy job, this same user experiences two other jams in the main processor unit, each time seeking key operator assistance. After the second jam, the key operator checks to see if the paper has been loaded correctly, a possible source of the jamming problem. The paper has been loaded correctly and she goes ahead and clears the jam. However, after the third jam the key operator decides to call a technician. Because the key operator received all three reports of jamming, she was able to construct a model of the machine's behavior and ascertain that the problem required the attention of a technician. When the technician arrives the key operator is able to provide information regarding the copier's recent history of jams. With this information in mind the technician quickly solves the problem and then reports to the key operator that the copier is ready to use. Not only was help readily available when the user experienced trouble, but because the report of trouble consistently went to one of the key operators, a relatively complete account of the machine's behavior was available.

In comparison at the low dependability site, a casual user experiences a jam in the main processor unit. At this site casual users are able to clear these jams because the machine is left unlocked. The casual user asks a passing co-worker, who is not a key operator, if he knows anything about clearing jams. The co-worker indicates that he knows very little and is too busy to help. The casual user walks away saying, "Oh, I'll use the other copier," leaving the copier in the jammed state for the next user. On this particular occasion two other people approach the jammed copier, commenting "Oh, it's jammed" and "It's still down," but not attempting to clear the jam nor asking for assistance. Finally, it comes to the attention of one of the key operators and she clears the jam. It later was reported that at least two other jams occurred that day with what appears to have been similar responses on the part of casual users. After each jam, the copier remained jammed for some period of time before someone, not always a key operator, cleared it.

Unlike what occurred at the high dependability site, no single person or group at this site consistently received the kind of information that would allow a model of the machine's behavior to be constructed. With such a model, the problem could be diagnosed and either resolved by a key operator or responded to by a technician. At the low dependability site a jamming problem might go on for a day or more before technical help is sought, all the while casual users are developing the perception that the copier is not dependable. Futhermore, when the technician finally arrives he has nowhere to go for an account of the copier's performance problems, because this information is scattered throughout the site. Likewise, casual users have no way of finding out about the copier's current status. The information needed by the casual users to decide if clearing the jam would result in their ability to complete their copying task is not available. Under these conditions users are likely to walk away from a jammed machine, which only exacerbates the problem.

User Anonymity. Related to the issue of consistent reporting of machine performance problems is the ability of users to remain anonymous. At the high dependability site a casual user experiencing a machine performance problem will find it difficult to walk away from the problem and expect his identity to remain undiscovered. Not only are the key operators, because of their proximity to the machine, likely to know who was last using the machine, but because there are relatively few users, approximately 30, locating the "responsible" person would not be particularly difficult. Also, as mentioned before, there are fewer reasons at the high dependability site to walk away from a problem since the key operators are nearby and willing to help.

On the other hand, at the low dependability site a casual user experiencing a problem runs little risk that his identity will be discovered if he simply walks away from the problem. No one is monitoring the use of the machine and identifying the "responsible" user would be practically impossible in an environment where approximately 150 people scattered throughout a five-story building use the copier. This increases the likelihood that a user will walk away from a problem instead of reporting it to a key operator.

The consequences of user anonymity are twofold. First, a single machine problem can cause problems for multiple users because the problem is never reported to a key operator who could either solve the problem or request the services of someone who could. Second, important diagnostic information is potentially lost when an anonymous user walks away from the problem without reporting it to a key operator.

Accessing the Community's Knowledge Base. Central to the issues raised so far is the ability of users to access the collective knowledge of the community of users. In other words, how does one gain access to the experiences and insights of other users? In the preceding examples when the key operator was in a position to act as a repository for information about the day-to-day operation of the machine, gaining access to much of the community's knowledge simply involved consulting the key operator. However, when no one was in a position to monitor the use of the technology, it was impossible to know where to go to find answers to machine performance questions. As such, the patterns of interaction among casual users, key operators, and technicians altered the ways in which the machine became integrated into individuals' day-to-day experiences, which in turn had consequences for users' evaluations of machine performance. At the high dependability site the structural position of the key operators relative to management and other causal users, the physical location of the machine, the number of causal users, and the responsibilities assigned to the key operators resulted in information about machine operability, performance history, and machine status being channeled consistently through the key operators. Consequently, casual users had available a ready source of important information about the copier. The key operators at this site were in a position to act as a repository for information about the machine.

On the other hand, at the low dependability site the structural position of the key operators within the organization, as well as the number of users, location of the machine, and distribution of key operator responsibilities resulted in the exchange of copier performance information that was sporadic and incomplete, with no single person or group acting as a repository for this information. Casual users for the most part were left on their own and because of this, many chose to walk away from machine problems when solutions were not readily available.

Figure 11.2 schematically represents differences in the interactional patterns at these two sites as they relate to the exchange of machine performance information. By observing the everyday activities of individuals, we

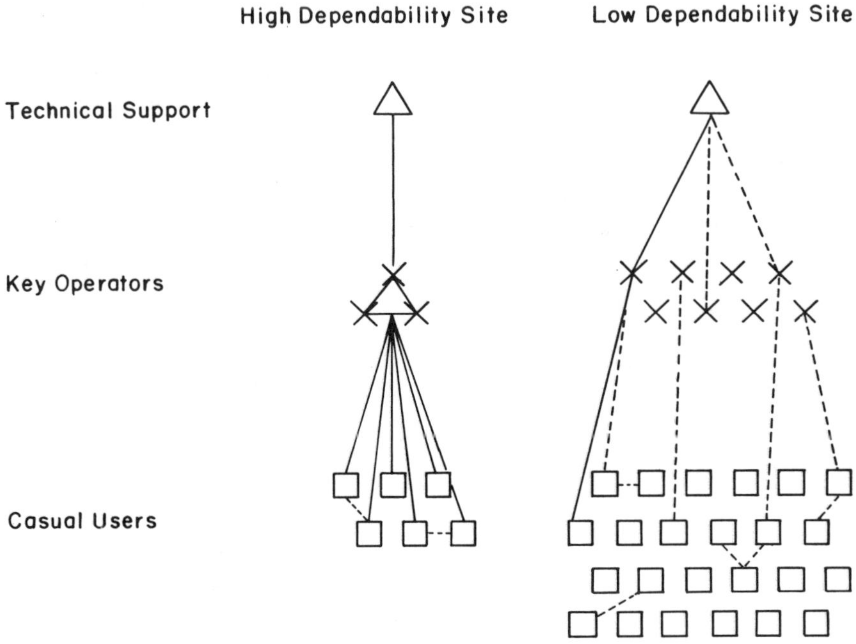

FIGURE 11.2. Information flow among technical support, key operators, and casual users.
____ = established, regular information flow
– – – = sporadic, incomplete information flow
Solid and dashed lines are based on field observations and not exact numbers.

can see how the social structural characteristics of the organization are related to patterns of interaction and, in turn, how these interaction patterns affect the availability and quality of machine performance information. In this way the users' experience of the machine becomes linked to social structural differences.

INTERFACE DESIGN COMMUNITY

The study of the use of Trillium, a computer-based design environment, by human factors and industrial designers provides some interesting parallels with the copier study. Trillium is used in the design of user interfaces for copiers, duplicators, and printers. The system allows designers to develop prototype designs for the user interfaces of products under development. It is no longer necessary to wait for the product to near completion before a functioning interface can be built and evaluated. Furthermore, designers who lack a computer programming background are able to design and build these interfaces. Trillium provides the designer with the means of creating and testing design concepts directly. For example, a designer can experiment with the pros and cons of using cascade buttons to specify a reduction or enlargement size or with different ways of giving the user a means of changing the contrast of a copy. Furthermore, the size, shape, and location of the cascade buttons or other physical components of the user interface can be changed easily. Although some design concepts used as building blocks in the development of user interfaces, such as number pads, sensitive regions on the screen, cascade buttons, and so forth, were included in the "original" release of Trillium, designers are able to add new concepts as the need arises.

The Trillium example also highlights the importance of having someone in a position to receive, organize, and disseminate information about system performance. As was the case with the copier example, the presence of a person to act as a repository for information about the operation of the system is critical to the successful integration of this technology in to the day-to-day activities of designers.

Social Structural Characteristics

The social structure of the Trillium user community is undergoing rapid change. The number of users is increasing and the uses for Trillium are expanding. The user community (including programmers who support the system) originally consisted of approximately six individuals all from one department within a larger organization. As of 1984, 40 people from four departments were using the system. When the user community was small, a designer with previous system experience became the focal point for the community. Designers and programmers channeled a great deal of information through this individual (focal member). However, as the community has grown this individual has not been able to function as effectively in this capacity. She now serves only a portion of the user community.

At the time of this study Trillium was relatively new and its place in the design process had not been fully determined. Management was unclear about what support personnel would be required to optimize the use of the new technology. The designer who was initially the focal member of the user community was not hired to perform this function. As the community grew, and consequently the time required to facilitate the use of the system increased, management was reluctant to commit this designer's time to these support responsibilities. The result was that some of the designers were left on their own to piece together the information they need to successfully operate the system. This has had some negative consequences on user satisfaction and the efficient use of the technology.

Role of the Focal Member. When the user community was small, the role performed by the focal member involved teaching designers, most of whom had little previous computer experience, how to use the system. In addition, the focal member was the main source for information about the new design concepts that were created by other designers and programmers. Making these new concepts available to other designers benefits everyone, because it limits duplication of effort and supports the standardization of designs. The focal member also assumed much of the responsibility for teaching designers how to use new features added to Trillium by programmers in response to designers' needs. In general, the focal member was the designer's main source for information about the operation and evolution of the computer system and the programmer's link to the designers.

When the community grew both in terms of the number of users and the number of departments using the system, the focal member was no longer able to provide the same level of support to all members of the community. In other departments using the system, no single person emerged to take over the responsibilities of the focal member for their group. Users with problems went to a variety of people, both designers with more system expertise and programmers, for help. This pattern of information exchange characterized most of the new groups using the system.

Interactional Patterns

Because designers went to a variety of people for assistance, with no one person responsive to the needs of the entire user community, it became more difficult to collect, organize, and disseminate system performance information. Furthermore, because information about system performance was distributed throughout each of the departments using the system, consolidating this information for later redistribution to the entire community of users was difficult.

Exploiting the Full Potential of the Technology. Those designers who were left more or less on their own to discover the capabilites of the system were less likely to be able to exploit the full potential of the technology. Without an overview of what the technology can provide, useful features of the system and more effective ways of using known features were overlooked. For example, a designer using Trillium was observed attempting to edit a portion of his design using a cumbersome editor not designed for such a task when a better editor was available but unknown to the designer. Had he been given an overview of the system or had someone been monitoring his use of the system, he would have known about the editor appropriate for the task at hand. In addition, an awareness of what other designers have produced using the system provides the beginning designer with some notion of what is possible. Designers who have not been exposed to the system often ask, "What can you do with it?" and "Would I be able to do such and such if I had access to the system?" The efficient use of the technology requires that there is some means by which designers can become exposed to the technology and to the experiences and insights of other designers. In the absense of a focal member responsible for supporting the user community, this becomes problematic.

Accessing the Community's Knowledge Base. The experiences and knowledge of each designer is but a subset of the experiences and knowledge of the community of users. The interactional patterns among members of the community are instrumental to the ways in which the collective understanding of the community is made available to individual users. When Trillium was first introduced, much of the system performance information was exchanged through face-to-face interaction and, as mentioned earlier, much of this information was channeled through a single designer. Users could be fairly certain that if they went to this focal member with a problem or question she would know how to access the community's knowledge base in her search for an answer. For example, if a designer was trying to get the effect of a flashing button on the interface, the focal member would be able either to direct the designer to an already existing "item type"[1] that represented this design concept, or pass the request on to a programmer who would be able to create such an item type. Without a focal member to act as a repository for system performance information, the designer might never become aware of the fact that such an item type existed and consequently might decide either to do without the effect or directly ask a programmer to create the item type. If the latter condition obtained, there would be two somewhat different item types that produced similar effects. The obvious problem with

---

[1] An item type is a piece of computer code describing both the physical and relational properties of a design concept.

this situation is that the efforts of the programmers have been unnecessarily duplicated. However, a more subtle problem that also follows from this situation is that the work of the two designers might not be compatible because each of their interfaces is dependent on a different set of item types. There is a need for some sort of clearinghouse for system performance information.

Sharing Design Innovations. A potential of this new technology that has not been fully exploited, partly because the item types employed differ somewhat among design groups, is the sharing of design innovations. The longer Trillium is in use the larger the collection of interfaces produced using the design tool. Trillium gives designers the ability to access electronically these interfaces in order to learn from the creative efforts of others. Design groups do not have to reinvent design concepts that have proven useful to others. In addition, exchanging interfaces and design innovations may lead to standardization of designs across product lines. Realizing this potential of the new technology will require that the role performed by the focal member be supported. The user community has grown too large for a single person to take on the responsibility, on an ad hoc basis, of creating and maintaining a data base of user interfaces along with their requisite item types. Relying exclusively on face-to-face communication is no longer viable because of the size of the community. However, some means of organizing and making available to designers the collective knowledge of the community of users remains necessary. It is important that both human and electronic means of facilitating this process be adequately supported in order to realize the full potential of the technology.

Electronic Support of Information Exchange

One of the advantages of this new technology is that it is adaptable to the varied needs of the designer. The designer does not have a fixed set of design concepts to work with, but instead can create new sets as the need arises. However, to the degree that designers can benefit from the ideas and work of others, they need to be able to keep abreast of the additions and modifications made by other designers. Accomplishing this objective in an environment where the user community is growing and where the technology is undergoing continual change is indeed a difficult challenge. The collection, organization, and dissemination of system performance information, a task previously performed by a human agent, may need to be supported electronically. However, designing electronic support for this function requires an understanding of how system performance information currently is exchanged through the day-to-day interactions of co-workers.

The objective in designing electronic support for the organization and dissemination of system performance information should be to supplement and facilitate existing patterns of information exchange. Designing electronic support without an awareness of and sensitivity to these patterns may result in suboptimal use of the support system and/or a disruption of existing, functioning patterns of information exchange.

CONCLUSIONS

In conclusion, if steps are to be taken to increase user acceptance and satisfaction, or to improve the effectiveness of the technology, it is important that the mechanisms by which new technologies become embedded in the workplace be understood. Although certain social structural variables may be associated with user satisfaction and acceptance, in order to make wise decisions either about "engineering" the social context of machine use or adapting the technology to support the interaction patterns, one needs to understand the embedding mechanisms. In particular, one must understand the nature of the everyday social interactions that serve to integrate the technology into the ongoing activities of the office or larger organizational unit. We need more studies that focus on how individuals accomplish their work in specifiable contexts.

12 Job Design and Quality of Working Life

James C. Taylor
Socio-Technical Design Consultants, Inc.

ABSTRACT

This chapter addreses two main points. First, new information technology has been introduced into white-collar work to automate routine. This means technology drives decision making, jobs are not being reconceptualized, and employees are alienated from their work. Yet a service occupation using automated tools requires employees to be dedicated and thoughtful exception handlers, rather than machine operators. Second, socio-technical design strategies are useful tools to bring together technology, organizational goals, and people to generate jobs that accomplish the organization's mission and contribute to the quality of working life—in particular those aspects of quality of working life usually associated with the professions. This theme is elaborated by examples showing how socio-technical design worked and by detailing the analysis tools that socio-technical design uses.

TURBULENT BUSINESS ENVIRONMENTS AND JOB DESIGN

American society, its people, its institutions, its business environments and its technology have all changed dramatically during the past 40 years. These turbulent environments, changing constantly in magnitude of force and direction, were identified 20 years ago by Emery and Trist (1965). Yet, in job and organizational design, many managers continue to copy from the past as if nothing significant has happened. They are reproducing organizations structured around principles and assumptions some of which were already old at the turn of the century. Rigid bureaucratic rules, narrowly defined jobs, centralized hierarchical control, and limited work authority—these industrial-era values are anachronisms today. More and more managers are reporting that the old ways don't work effectively. They recognize the need to design new work systems that promote flexibility and commitment, but they lack the knowledge of appropriate models to guide them in this task. Some search in vain for packaged solutions, for the magic elixir that will guarantee a quick fix without disturbing the status quo. Others realize that these "problems" also can be seen as opportunities, and are working to develop long-term programs that involve employees and their representatives at all levels in consciously shaping their future.

Performance Improvement and New Technology. Apart from replacing personnel with machines, office automation seems to be failing to deliver measurable improvements in effectiveness. Most successful applications of office automation show evidence of cost containment, but are unable to produce quantifiable results proving accelerated enhancement of end-user performance. I assert in the remarks to follow that new technology can aid in improving organizational performance, but only by that technology being integrated into the organizational system and into the human aspects of work. Current applications have failed largely in this regard not because of the technologies themselves but because they have been misapplied, as guided by the industrial era values. Those values guide managers to apply technology to continue to mechanize old tasks, rather than to develop new tasks that are consistent with organizational purposes and human values. Are we as a nation still in the business of producing high volume, unchanging products like the Model T Ford and the 3% interest passbook account of the industrial era? Conversely, are we really serious about being in the business of delivering highly variable, high quality products like microcomputers, and services like money market accounts, and universal life policies suited to ever-changing customer needs? If indeed we are engaged in a flexible response to changing markets and environments then simply replacing people with automated machines (however flexible they may be) will not prove successful.

Jobs and Automation. Automated technology absorbs routine activities, segregating them into machines. As it does this, it creates new relationships between technical processes and the employees who are involved. People who work in automated systems often function as interdependent actors who respond appropriately when something goes wrong. Sometimes, this is by virtue of the design of jobs and organization; but often it is unknown to the managers of the enterprise. Often the real role of employees in a highly complex technological environment is revealed during times of labor–management conflict; for instance, air traffic controllers exercise considerable discretion except during the work slow-downs in the late 1970s when they meticulously followed all work rules. Management called this behavior "malicious obedience."

Examples of this management dependence on employees in advanced technology applications are widespread and often are observed in less confrontational settings. As evidence of the diversity of type of work and situations, the following paragraphs illustrate management dependence on operators in a mainframe computer center, on secretaries in an automated engineering office, and on workers and supervisors in a new manufacturing resources planning system.

A socio-technical design project with a large computer operations center revealed that despite well-meaning individual managers the center was founded on (and managed with) values that emphasized that machine resources were crucial and the human resources were routine and replaceable. Operators in this and similiar centers were referred to as "tape jockeys," "tape monkeys," or worse. As the project in this center progressed the operators chosen to take part in the analysis and design described their work as crucial to the success of the center because of their attention to subtle distinctions in the requests and instructions of the center's users or customers. These operators spoke of "detective work" they undertook to accomplish what the users wanted, without delaying the job by bothering the user with questions of clarification. The operators were proud of this discretionary activity, but they wished that management understood more of what their work entailed. As the redesign project reached its recommendations management did become aware of the contribution made by the operators and chose a design that encouraged and rewarded operator initiative and discretion (Taylor, 1981).

As a second example of automation and dependence on low-level employees, a large French engineering firm, SERETE, has described its own experience with multifunctional engineering workstations, and what it refers to as the role reversal between engineers and their secretaries (Audibert, 1984). At SERETE the traditional relationship between engineers and secretaries evolved toward a new one when engineers requested the engineering workstations, but complained about the edit time required to complete their documents. These complaints led to training secretaries on the multifunction stations and training those engineers, who requested it, on word processors. Audibert reports that optimum profitability seems to be achieved by giving engineers a workstation that allows them to enter their own text easily, and permits later transfer of this text to more sophisticated [multifunctional] workstations for page setting, printing, inserting diagrams, and so on by secretaries. Clearly, SERETE relies increasingly on its secretarial staff for the effective operation of its engineering workstations, and it finds engineering performance thereby enhanced.

A third example shows how real world complexity defeats the effective application of modern computer-based office and manufacturing control systems without the informed and committed intervention by workers and supervisors. In a recent study undertaken for the U.S. Office of Technology Assessment (Schneider, Howard & Emspak, 1985) a number of specific examples of management dependence on workers and foremen in a Manufacturing Resouces Planning System (MRPS II) were described. These authors state:

> despite the sophistication of MRPS II as an information system, as a production planning and control system, it is often too 'rigid' or too 'simplistic' to adequately reflect the complexities of production on the shop floor. As a result, the managers and workers in the [shop], as well as the supervisors and clerical personnel in the Production Support office, all have to devise ways to 'end-run' around the system, simply in order to get the work done. Or, as one dispatcher from the [shop] puts it: 'If you want to keep people working—and they do want to work, believe it or not—you have to cheat the system. . . .' [We are] witnessing the creation of a new informal system, in which the role of the worker is to handle those frequent situations where the computer system, however technically sophisticated, is too crudely programmed to respond to the complexities of the real world. (pp. 50–51)

For economic reasons as well as service quality, unexpected problems need to be corrected as quickly as possible; and this imposes three new requirements on those who do the work. First, workers need a large repertoire of responses, because what they could be called on to do is not always known in advance. Second, they cannot depend on or be controlled by supervision in the traditional sense, because they are required to respond immediately to events that occur with little or no warning. Third, they must be committed to doing whatever is necessary on their own initiative. Thus, organizations using automated technologies are highly dependent on the motivation of their employees if they are to succeed. Economic goals will not be met unless the work process functions properly and this will not happen unless appropriate actions are made in response to (and in anticipation of) unexpected events. These appropriate responses will not be undertaken unless people are committed to their functions. Commitment cannot be bought or forced, but can only arise as a result of people's experiences in their working situation.

The Industrial Era and Job Satisfaction. A new approach to job and organizational design raises the question: What aspects of working life do organizational members consider important?

Under industrial era assumptions that question is ignored in favor of the question: What characteristics should job holders possess in order to conform to or "fit" the demands and requirements of the present technical system?

Historically, industrial psychologists have advised managers to hire individuals whose skills and personalities matched what was seen to be the fixed and unalterable characteristics of work. Since the publication of Hoppock's (1935) job satisfaction survey, the fit between person and job could be measured using standardized indices of job satisfaction. By the 1970s, when employee absenteeism and turnover rates had grown to all-time highs, surveys reported the astounding statistic that 80% of American workers were, at least, not dissatisfied with their jobs (Taylor, 1977). This is in part a result of the selection "fit" just described; together with widespread acceptance that when it comes to work, the technology counts more than the people. As previously noted, technology requires new worker behaviors and attitudes. Traditional job satisfaction research has done little to reveal what workers want from work.

THE QUALITY OF WORKING LIFE

A constructive approach to the problem of incorporating human preferences into the design of organizational systems is offered by quality of working life research. This approach consists of identifying those aspects of work that are seen as desirable by organizational members, and incorporating that information into initial design considerations. This position is realistic if we understand that today's computer-based technology is so rich in potential variations and arrangements that design decisions can depend almost exclusively on the social side of the equation (Davis & Taylor, 1976). For instance, early highly rigid word processing centers were a source of operator alienation and frustration. With high operator turnover and low operator commitment, these centralized units could not be considered successful (Taylor, 1980; Williams 1981). However, with time and the development of a variety of social organizations in word processing, the variable of centralization–decentralization has been found to be independent of the innovative and effective use of word processing technology (Johnson et al., 1985). The effective design of jobs and organizations requires that the construct of quality of working life be specific and concrete as defined by organizational members. Levine (1983a, 1983b) reports some recent research into comprehensive employee-generated lists of quality of working life that reveals that company-specific measures developed by his participative methods are as much as 50% more powerful as predictors of employee attitudes than more generic instruments such as the Job Diagnostic Survey (Hackman & Oldham, 1975). Further, these comprehensive employee-generated lists reveal that employees value a job that is central to the essential business of the enterprise, and that this centrality is an important contributor to employee attitudes when job design enhances it (Gustavson & Taylor, 1984; Levine, 1983b). But these studies also found that it is not enough to be central—employees must also feel competent at what they do, and they must feel that the product of their work and the company's purpose are important and relevant to their community or the society (Emery & Emery, 1978; Levine, Taylor, & Davis, 1984). These three aspects of quality of working life—centrality, competence, and relevance—are powerful and newly recognized sources of employee commitment and organizational effectiveness. They occupy an important place among the set of items already acknowledged as contributing to high quality of working life (cf. Biderman & Drury, 1976; Hackman & Oldham, 1975).

INNOVATIVE USE OF OFFICE TECHNOLOGY

In 1982, several of us undertook a study of the innovative applications of new office technology. Two hundred organizations were surveyed to learn more about the conditions under which technical adaptation occurs, and the effects of such innovation on organizational effectiveness (Johnson et al., 1985). Word processing was chosen as the technical focus of the study because as a reasonably mature component of new office information systems, adaptation could be expected. We were interested in adaptations that visibly contributed to organizational productivitiy or effectiveness. Examples of such adaptation include: Experimenting with "what ifs" in written narrative much like spreadsheet programs permit experimenting with numerical information; using the find-and-replace function to perform key word analyses of reports during preparation to bring the cycle of complex document generation into "real-time"; discovering ways to readily format unusual or complex information arrays such as scientific or mathematical expressions, multi-columnar layouts, and the like.

The focus of the inquiry was on how management decisions shape use of the technology. The results of that study describe a technological innovation in transition. The study revealed that word processing is a new but rapidly maturing technology. Of our sample, 61% of companies had used word processing for 5 years or less. A primary finding revealed that only 32 of the 196 companies responding (some 16%) could say that they were making innovative and effective use of their word processing equipment. The other 84% reported that they had not discovered effective new uses for their equipment, and most were using the equipment merely as more highly mechanized or efficient typewriters. Although many organizations could report innovations, most of these were either trivial (e.g., having a Videc machine play a few notes of the "Star Spangled Banner"), or isolated in "pockets" of individually clever operators.

We found that most word processing planning has been done in terms of equipment procurement (contemporary studies confirm this, as noted by Bikson, and Bjørn-Andersen in this volume). Implementation has been left to independent supervisors responsible for training and development of applications. The initial trend was toward centralization of equipment. With lower costs and increased user friendliness of equipment, much of the rationale for centralization (e.g., continuous use, operator training) are not as valid as they were in the late 70s. A terminal on every desk is now a reality in many work units we studied. Most organizations with word processing centers clearly benefited from them; in many, the introduction and implementation of word processing might not have been possible any other way.

Many of the organizations we studied are moving toward distributed word processing and broader accessibility to word processing equipment, with authors (e.g., engineers, auditors, lawyers, purchasing agents, teachers, managers) writing and editing on line. We found some evidence to report that such professionalism seems related to a diminution of specialized clerical jobs such as word processing operator, and the beginning of new technology-based roles such as "assistants to" managers or professionals where the former clerks coordinate closely with their professional collaborators to provide a wider range of direct technical support (Johnson & Taylor, 1985) . This is similiar to the role reversal between engineers and secretaries noted by Audibert (1984). Despite this "professionalizing" of word processing, its organizational planning remains focused largely upon equipment procurement rather than upon people or jobs. This is elaborated upon in the discussion to follow.

A second data collection was undertaken among a smaller second-wave subset of the original sample. This second wave involved field visits and questionnaire surveys among word processing supervisors, operators, and authors who are customers of word processing services. All of the 32 sites reporting innovative and productive uses of word processing during the wave-one survey were included in wave two, and were matched with a roughly equal number of sites reporting no innovative and productive use of the technology. Table 12.1 shows the distribution of sites visited for the wave-two data collection and some of their characteristics.

Of the organizations that reported that they were making innovative and effective use of their word processing equipment ("innovative" sites) there were about equal numbers of centralized and decentralized sites (see Table 12.1).

Of the innovating and effective distributed word processing systems (see Table 12.1) most of the managers, operators, and professional users could describe the purpose of word processing within the larger mission of their company. They had a "systems consciousness" that permitted them to effectively exploit their new technology. They knew what business they were all in, and they knew how word processing helped them achieve it. They also reported that the innovations developed were guided by their mission. Professional users in these sites commented that word processing technology was "effective" only insofar as it helped them in their professional activities. The

TABLE 12.1
Four-way Classification for WP Innovation

	WP Structure	
	Distributed Sites	***Centralized Sites***
Innovative Systems	"High Integration Systems" (n = 14) *uses:* Adaptation as a way of life, focus on professional augmentation *monitoring:* Focus on increased capability *diffusion:* Organized communication of new ideas (often through Email). *management:* Considerable communication about WP and adaptation. Much operator participation in decision making.	"Expanding Systems" (n = 18) *uses:* Serving professionals, evolving WP applications. *monitoring:* Service quality, and new applications. *diffusion:* Through supervisor salesmanship. *management:* Extensive communication about WP and adaptation. Little operator participation in decision making.
Adopting Systems	"Low Integration Systems" (n = 14) *uses:* Electronic typewriters, little automation of routine. *monitoring:* By hours of WP use. *diffusion:* "Pockets of cleverness" remain isolated among WP operators. *management:* Unrealistic expectations about efficient use of machines. Weak or rigid supervision, little participation or comunication.	"Clockwork Systems" (n = 15) *uses:* Batch typing, boilerplating, no ties to professional users *monitoring:* By lines/pages of output for total unit. *diffusion:* Limited expansion of variety of application among WP operators, or with authors. *management:* Emphasis on volume and efficiency, little communication about adaptation, little participation in decision making.

productively innovative centralized sites had word processing center supervisors with a clear systems consciousness. These supervisors were very service-oriented and spent much of their time in contact with authors, either determining author needs, or selling word processing innovations intended to address those needs. In some sites supervisors "protected" their word processing operators from authors, whereas in others, supervisors encouraged interaction between operators and authors. Both supervisors and operators in both types of innovative sites reported the conditions for innovation included making time available for inventing, and encouragement of innovation for customer needs as well for less specific requirements. Supervisors in the adoption-only sites reported an emphasis on disciplined efficiency (adopting/centralized) in lieu of innovation, or casual ignorance of the potential of word processing (adopting/distributed). This ignorance is reflected in the fact that merely spreading word processing access to professionals does not relate to effective innovations in word processing. Most of the adopting, distributed sites were among the worst examples of effective use of the new technology; with the dominant attitude being that word processing is as necessary as pencil sharpeners and to be procured and used with about the same zest. Both the centralized and the distributed adopting sites were rated as less effective by both their members and their customers/users. Word processing respondents from all four types of sites reported a strong interest in being flexible and innovative to help their companies.

A major conclusion of the study is that responsibility for management of equipment and innovation in applications cannot continue to rest with the entrepreneurial word processing center supervisors. Technological trends are that word processing will become increasingly distributed, with the replacement of dedicated word processing equipment, with more flexible and comprehensive personal computers and office information terminals. The innovative use of word processing, electronic mail, telecommunications, database management, and other functions will require continued support when equipment is broadly distributed, but those people we found to be providing most of the guidance will not be organizationally able to do so without real changes in their organizational position and in management policy. As the findings emphasize, people responsible for operating the equipment often are most interested in adapting the technology to the work to be done. Their quality of working life is enhanced by contributing to organizational effectiveness through innovation and competence. These people, however, often lack perspective of overall organizational structures and mission. They are not always in close touch with the authors, their customers. They also usually lack the time and management encouragement to experiment, even if they are conscious of their employer's enterprise. Systems planners and other computer specialists can help provide the organizational perspective but often have no real sense of daily operating problems and potentials. Our research also showed that the word processing managers and supervisors have a higher opinion of their subordinate office workers than systems planners do on average (Johnson et al., 1985; Taylor, 1979).

TECHNOLOGICAL DETERMINISM

Scratch any piece of technology and you will find the values and aims of a society it was designed to serve. For technology is like genetic material—it carries with it the code of the society that conceived it. This is why technology "determines" skills, jobs, social relations, and organizational structures

(Davis & Taylor, 1976). What are the values of technical systems designers, and how are those values shaped? Since the 1950s environmental complexity and turbulence has increased (Emery & Trist, 1965), not diminished, and the demands on work systems to be flexible have increased exponentially. Despite this environmental pressure, I have found that job designers (systems analysts as well as industrial engineers) still emphasize machine-like efficiencies over any value of unique human contribution (Taylor, 1979). In that study, over 100 production engineers, engineering managers, systems analysts, and data processing managers in California were asked to list the five most important principles of job design. They identified maximizing operational cycles per unit time; efficient use of machine resources; making jobs as simple as possible; reducing manpower; and providing management with better information. The principles are those of the industrial era, and the designers tend to explain their use of them as a matter of following the directions of their firm's top management. Because today's managers continue to tacitly take these principles as guidance, design proposals for today's work systems including advanced office information systems often are based upon outmoded concepts. Because this bias is so pervasive and so little acknowledged, proposals that rest on other principles, such as socio-technical systems, which I describe later, are viewed suspiciously.

THE PURPOSE-ORIENTED APPROACH TO WORK DESIGN

Organizational improvement and job design efforts can be said to fall into three categories. All three embrace the general rubric of "organizational development" with a specifically organizational focus. The three major types are the problem-oriented, solution-oriented, and purpose-oriented improvement efforts.

The problem-oriented approach of organizational improvement and effectiveness states "Let's deal with what's wrong with the organization—let's find out what the problems are and address those." It deals mainly with today's (yesterday's?) problems, and it doesn't really look much to the future. It does not design for organizational purpose. One enduring example of the problem-oriented approach is the organizational survey. Basically, this approach involves doing a large scale paper and pencil survey and trying to find out what the organization's weaknesses and job dissatisfiers are, and then working on those weaknesses. If the best predictor of tomorrow is yesterday and, with a reasonably placid world, survey feedback is an adequate way of approaching organizational improvements.

The second organizational improvement approach is that of solution orientation. Examples of this approach are programs like job enrichment, office automation, and quality circles. Included here are techniques that will improve organizations, even if they are already doing "ok." There are many techniques that we could include in the solution-oriented model; better accounting methods, better production planning, and new technology could fit in that category. These solution-oriented approaches are characteristically faddish, and they are not specifically designing for the organization's purpose.

With the third, or purpose-oriented approach to organizational improvement, managers take the stance: "Let's re-evaluate what business we're in and then consciously plan to do that business as well as we can." This stance does not mean a broad choice or distinction of "business" ranging between say, oil drilling and manufacturing office equipment. Rather a more subtle distinction is intended between what business we might be in if we just continue to float along with our market, customers, and competitors; or if we chose, to be distinctively different within that marketplace in a way that captures our imagination and excels in that marketplace. Examples of this subtle difference in mission orientation are the purpose or mission of a dedicated teaching hospital compared with the hospital providing high quality community health care or the hospital delving deeply into disease research.

I recommend the purpose-oriented approach over the other two more familiar approaches when planning for the organizational integration of new technology. I do this because the purpose-oriented approach focuses on the long-term aspects and on the purpose or mission within turbulent and dynamic environments. I also recommend the purpose-oriented approach because it is proving successful in applying technology in the 1980s.

SOCIO-TECHNICAL SYSTEMS DESIGN

The ideas of designing for environments and purpose are not new. For the past 30 years Drucker (1959) and Trist (1981) can be seen, in their separate ways, as the conceptual fathers of purpose-oriented organizational improvement. Socio-technical systems design is the best known of the purpose-oriented approaches. In this context, socio-technical systems is a methodology for improving organizational processes and structures through discovering what business we're really in; and what joint optimization between social and technical requirements are needed in order to excel at that business (Davis, 1981; Emery, Foster, & Woollard, 1978; Taylor, 1978). This is opposed to simply saying socio-technical systems is a managerial technique of applying group solutions, job enrichment, or office automation as ways of managing or of solving organizational problems.

Socio-technical systems is a high-level management approach for improving any work process in a manufacturing or service organization. It has been developed over the past 30 years in Europe and the United States, and the results of the success stories set new world-class standards for manufacturing effectiveness.

For example, Proctor and Gamble began using the approach as early as 1969 and has continued to refine it over the past 15 years. Zilog, Inc., a subsidiary of Exxon Corporation, applied socio-technical systems management to the start-up of its Nampa, Idaho, semiconductor plant in 1979 and revolutionized the way microprocessors are manufactured. The plant reached full production in 13 months instead of the more typical time of 25 months, circuit yields were improved by 100%, and employee turnover was reduced to only a few percent per year.

Why Socio-Technical Systems Management Works

There are three reasons for the success of socio-technical systems management: joint optimization, purpose orientation, and structured process.

Joint Optimization. Socio-technical systems management gets its name from the way it integrates the production (technical) requirements of the work process with the organizational (social) functions of the people working in the process. The term *work process* is used here because socio-technical systems management can be applied to any process in which work is being performed. These processes include manufacturing production, transaction processing, research and development, engineering, data center operations, and clerical operations.

Many industrial engineering techniques focus on the efficiency of the work performed, which is the technical part of the work process. Though the industrial engineering practitioners consider social functions, they do not integrate them into the management of the process in any systemic way. The "people" portion of the process usually is organized around the technology on the basis of seat-of-the-pants opinions of how best to manage people. Organizational development techniques, however, focus more on the social functions in the work process, but the organizational development practitioners tend to overlook the work that needs to be done.

Socio-technical systems uses many of the industrial engineering and organizational development principles, but the focus is on integrating them to obtain joint optimization of the social and technical functions in the work process. The work process that optimizes the technical functions often is not optimal for the social functions, and vice versa. What is best for the work process as a whole may be at a point that is not optimal for one or the other subsystems.

Purpose Orientation. Because socio-technical systems management focuses on the work process as a whole rather than on its parts, it is a systems approach. It starts with the particular mission or purpose of the organization and develops a design and system of management tailored to that purpose. This purpose orientation is much more powerful than the more common problem-oriented approaches (e.g., technical audits and questionnaire surveys) or solution-oriented approaches (e.g., office automation and quality circles), which address only a part of the system and which often ignore the strategic management issues and the overall mission of the process.

Some practitioners view socio-technical systems strictly as an organizational ideology or design technique. The real value of socio-technical systems is as a management system that functions as the operational or implementation portion of the strategic management process. Socio-technical systems management translates strategies into the missions needed to excel strategically, and it focuses on the effective implementation of those missions. Furthermore, it serves to refine and strengthen the operating missions, which in turn help to refocus the strategies of the company.

Structured Process. The third reason for the success of socio-technical systems management is the structured approach it uses to analyze and implement the operational improvements. Many management models are conceptually interesting but operationally soft. McKinsey's 7-S Framework is an example (Peters & Waterman, 1982). Most managers would find it difficult to implement such an approach because there is too large a gap between the concepts in the model and actual day-to-day activities in the work process. The socio-technical systems approach provides a process for data collection and analysis, and thus helps makes models like the 7-S Framework operational. Conceptually the 7-S Framework and socio-technical systems management are very similiar. However, socio-technical systems provides managers with the substance and structure that have been operationally proven over the past 30 years.

DESIGN AND REDESIGN OF ORGANIZATIONS

Socio-technical systems is a continuous management process involving four steps: (a) definition of the purpose (mission and philosophy); (b) analysis of the technical and social subsystems; (c) specification of the design recommendations; and (d) implementation of improvements in the work process. The social analysis builds upon the technical analysis to identify ways that key variances in the conversion process can be better controlled by improving the fit between the technical functions and the social functions. With the information developed in the analysis, managers and employees can implement improvements in the work process that: (a) identify and control variances at their source; (b) place coordination of the work at the lowest organizational level compatible with the technical and social systems; and (c) develop organizational structures that fit the structure of state changes in the throughput. It has been found with socio-technical systems management that high quality of working life is the result of being central to and competent in creating a meaningful product. Further, this competence and centrality is greatly enhanced by designing work so that employees consciously control key technical variances as close to their source as possible.

Socio-technical systems management is an iterative process. The analysis continues after improvements are implemented, resulting in a continuous process of analysis and change.

APPLICATION OF SOCIO-TECHNICAL SYSTEMS TO PLANT AND OFFICE

Socio-technical analysis and design involving new plant installations have been used in a variety of manufacturing settings. A few of the highly successful new manufacturing sites in North America include the Gaines pet food plant in Topeka, Kansas; General Motors plants in Fitzgerald, Georgia, and Brookhaven, Mississippi (Walton, 1979); the Shell chemical plant in Sarnia, Ontario (Sullivan & Davis, 1980); the Sherwin-Williams paint plant in Cleveland, Ohio (Poza & Markus, 1980); the Skippy peanut butter plant in Little Rock, Arkansas (An American miracle that works, 1982); and the Zilog semiconductor fabrication plant in Nampa, Idaho (Gustavson & Taylor, 1984). The Zilog case is especially interesting here because white-collar applications in the sales and engineering departments followed initial successes in the fabrication plant. Those three cases of socio-technical systems design application in Zilog are described now.

The Manufacturing Plant

In 1979, Zilog (a Silicon Valley subsidiary of Exxon and creator of the well-known eight-bit microprocessor, the "Z-80") was preparing to open a new semiconductor fabrication facility in Nampa, Idaho. Zilog had hired a group of 10 managers who had been involved in start-up plants in other companies to start and run the plant. They had all experienced the high level of frustration, the lengthy period needed to bring operators and process engineers to productive levels, and the high turnover of personnel involved in start-ups. They wanted to do this one differently. They also wanted to meet or exceed the productivity and quality standards set for the world in their type of product by the Japanese.

In the summer of 1979 Exxon and the Zilog president sponsored a 4-day retreat for some 30 Zilog managers from the president through unit managers three levels below. The twin purposes of this retreat were to expose these managers to new management and organizational ideas, such as open socio-technical systems management, and to begin to apply these ideas to the design of the new manufacturing organization in Idaho. Nine of the 10 managers slated to take part in the Nampa, Idaho, start-up attended the retreat.

Following the retreat the Nampa managers engaged an internal consultant (Paul Gustavson) and an external consultant (myself) and undertook to design their plant organization based on the new principles. Through the process of socio-technical analysis they clarified a mission for the plant that permitted members to focus on the purpose of making high quality microprocessors, the product of their efforts, rather than merely the process itself. They also created a philosophy that emphasized values such as self-reliance, teamwork, and mutual trust. With the consultants' assistance, the Nampa managers undertook a socio-technical systems analysis and design (the details of the socio-technical systems techniques are elaborated in the white-collar cases that follow). The socio-technical analysis in the Nampa plant revealed the patterns of key technical variance control necessary for successful semiconductor manufacture. In viewing this pattern of variance control the Zilog managers concluded that the large number of roles responsible for variance control in the typical plant in their industry created the chaotic disorganization they had all observed and experienced in previous jobs in the industry. In seeking to simplify this organization and to streamline relationships, the Nampa managers selected the manufacturing operators as most likely to be close to and capable of dealing with key technical variances as they occurred. The socio-technical analysis guided them to examine the necessary and expected social behaviors between those operators and the rest of the plant. For instance, operators were expected to perform routine maintenance on the equipment they used, thereby eliminating unnecessary breakdowns and delays. The role of the maintenance employees then became one of training the operators to do routine maintenance and of "hot rodding" the equipment so that it could perform more effectively than originally designed. The result of this planned change of fit between roles and technology would be improved equipment utilization, and equipment life that extended well beyond industry norms. In the Zilog Nampa plant the jobs and organization were designed to optimize the production of high quality products as well as to supportively challenge employees, to permit them to develop and advance and to involve them in the management process of building good microprocessor circuits.

By October 1979, the plant was hiring operators (who were called "manufacturing technicians"), and by January 1980 the first good circuits were coming off the line. This was phenomenal progress for start-up, and unprecedented in the industry. The work teams responsible for the production were specifically oriented toward the control of key manufacturing variances, and toward the mutual support and advancement of all team members. They were organized around stages in the product transformation instead of being organized by the technical processes of semiconductor fabrication, which is typical in the industry. The resulting design and its evolution during the past 6 years confirms the power of purpose-oriented design. The plant has retained its initial high levels of quality and cost performance while adapting its structure and processes to changes in the product market, to changes in corporate management, and to changes in technology. For instance, the phenomenal start-up performance for the first module was repeated with the addition of a second module of the plant to produce a more complex product with a state-of-the-art process. Funds for the second module were approved in December 1983. It was built, technicians were hired and trained and the module was producing good products by November 1984. Furthermore, personnel turnover, always a chronic problem in that industry, was addressed directly by the original design, and it has remained much lower than in the rest of the industry. In short, employees like it there and want to continue to be part of a high performance system. Ironically, it has been important for Zilog to ignore the specific design of successful organizations such as Nampa. Instead, the managers in that company remind one another that the socio-technical analysis and design process and the guidance of the design principles (in particular, a clear mission and philosophy) are key in creating the organization which is "right" for those people.

Customer Service Department

By May 1980, the vice president of Zilog's sales and customer service decided to apply socio-technical systems in improving his operation. He had been initially reluctant to participate in the 1979 retreat, but had found it a useful approach. As the year progressed he became increasingly impressed with the first-year successes at Nampa. To develop further interest in socio-technical design he invited the managers of his customer service department to visit the Nampa plant. After 3 days on site they were convinced that the socio-technical approach could help them improve their operations. They in turn suggested that the supervisor of the order entry function (reporting at that time to the financial vice president) learn more about the organizational innovations at Nampa, because order entry would be a critical function in improving the performance of customer service. The sales vice president then arranged to have the customer service and order entry managers serve as a design team for customer service, and appointed the Nampa consultants to assist them. A description of the steps in the socio-technical analysis and design of customer service follows.

Systems Scan. In July 1980, the customer service design team began the socio-technical analysis with a systems scan to develop an overall picture of that unit's work system and its environment. The environment for customer service included the production control function, purchasing department, accounts receivable, sales, and field service units. The customer service mission was developed. It emphasized order and product information services to customers and sales with maximum accuracy and timeliness and for low cost. The design team created a management philosophy similiar to that used by the Nampa plant employees. The design team also listed some problems in the existing customer service and order entry organizations, which included uncoordinated product introduction, delays in booking orders, and accepting and shipping orders over customer credit limits. Staff morale was said to suffer from a lack of management support and inequities in compensation.

Technical System Analysis. The customer service design team examined the conversion or transformation process of their organizational system. Their output or "products" were identified as "information for shipping accepted and booked orders," and "information to satisfy requests of customers and Zilog sales force about orders in process, returned orders, and new Zilog products." Their input or "raw materials" to be converted into output were identified as "customer requests for products (orders)," and "customer and sales force requests for information about price, delivery, returns, and information about new products." The design team next identified the state changes in their inputs as they are transformed into the outputs just described. These state changes, or "unit operations," included: (a) an answered inquiry for price or delivery information; (b) a received request for product delivery; (c) an accepted request; (d) a standardized order for the product request; (e) an acknowledged (booked) order; and (f) the confirmed shipment to customer. Having listed these essential state changes in their conversion stream, the design team identified 36 "variances" or conditions across the six-unit operations of the product stream that can vary during transformation. These variances in the first unit operation for instance included the source of an inquiry (customer, distributor, Zilog field sales representative, etc.), the urgency of of any inquiry (needed now or later, and how much later), and the time required to answer an inquiry (including length of delay for less urgent requests). The next step in the technical analysis was to identify the relationships among the variances. Thus, variances upstream that interfere with the control of variances in later unit operations were pinpointed on a graphic display called a *variance matrix.* Finally, the design team identified the most critical variances, or key variances, as those that must be controlled if the conversion process is to be successful in pursuit of their mission. Specifically, they found that delays in responding to inquiries was a key variance because of its relationship to the accuracy of the answers the department could provide; and also because of the inconvenience caused to customers and possible loss of business. They identified the quality of incoming orders as key because of the potential for errors in booking or shipping, and because of delays incurred in insuring accuracy. The design team also listed as key variances: time delays in orders shipped caused by obtaining support documents, licenses, etc.; the terms and conditions of payment; and whether the order was for special or standard products.

Once key variances were identified, the technical analysis was continued in order to describe the manner in which each of the five key variances was controlled. For each key variance, the design team listed the unit operations where each key variance occurs, is observed, and is controlled; the role (or roles) that currently controls it; the activities required to control it; the information and sources of information important in guiding control activities; and social and technical enhancement that might improve its control. This analysis of key variance control proved useful to the customer service design team in identifying the most crucial aspects of their conversion stream. In a later report on their technical analysis the team said, "As we continued through [the variance control table] we clearly began to identify our real key variances, and . . . we were able to focus on one in particular. This 'Key' Key variance is identified on the Matrix as 'Quality of Request': Translation, if the order we receive from the outside world has all necessary information, credit approval and supporting documents then the processing is smooth and the customer is easily pleased. This Key Variance is by no means easy to control. As a matter of fact we had so many suggestions for redesign and improved technology that they were too numerous to list [here]." In that analysis the design team noted that both customer service representatives and order entry clerks were important roles in working with the sales force and the customers to effectively control the key variance of "quality of request."

At that point in their analysis, instead of concluding their design efforts with a suggestion to maximize technology or to issue edicts obliging customers to submit all possible forms and information, the design team proceeded to scrutinize the social subsystem of the workplace to achieve the best match between technical and social factors.

The social analysis consisted of a series of inquiries by the design team to assess the expectations of the customer service and order entry clerks, and those of the relevant roles in the environment of their conversion stream, and of the customers and sales force. The design team's major findings were that adequate coordination could not take place between customer service and order entry as long as they remained in separate departments and separate groups. Coordination was expected by customer service of order entry, but it was not forthcoming. Further, the team noted that the sales force was largely out of touch with customer service and was never in contact with order entry. Finally, the pay differentials between customer service "reps" and order entry clerks was reported to be a point of friction between the two groups and acted as a tangible symbol of a difference in status.The socio-technical analysis was used to explain and justify the merging of these two functional groups that had a strong interdependence to complete the larger task they both shared.

Work System Design. The team recommended a high performance work system that organized tasks so that variance control, coordination, cooperation, and psychological job criteria enhance one another. The resulting design favored an organization based on work groups in place of the rather tall hierarchy of the then current organization in customer service. These work teams would manage a cluster of tasks and variances that result in some tangible output. These teams were organized geographically. Each sales area was assigned a customer service team that would be responsible for order processing and answering inquiries for that entire region. Each team possessed all skills (including order entry and customer contact) needed to provide full service. A sales support person from each regional sales office was designated as a member of the appropriate team, as well as remaining a member of the regional office. This person was given responsibility for coordinating activities with the region.

By February 1981, (about 9 months after the analysis began) the recommendations had been approved and implemented. Training sessions were conducted to teach the skills needed for working on the teams, and the teams began to function. Initial results showed improved quality of working life and better performance in a number of areas. Notable improvements occurred in turnaround time, error rate, level of service, general atmosphere, level of stress, and coordination with the sales office. The existing technology base of the customer service function was not heavily computer-based, and few improvements were made in it.

Design Engineering Department

Introducing new technology into service and professional organizations poses an added design challenge to managers, and socio-technical systems design proved to be effective for Zilog there as well. Taylor, Gustavson, and Carter (1985) describe the case in which Zilog design engineers and other employees were involved in the redesign of their own work to accommodate the critical role of a new computer-assisted design technology and the emerging technical skills of computer-assisted design operators in the increasingly complex design of new integrated circuit products.

Zilog's president and chief executive officer had attended the 1979 retreat previously described, and his interest in innovative organization and his concern for the value of human resource was well known. By late 1979 his interest in diffusing the organizational innovations that led to the successes of the Nampa plant to the company's design engineering function was becoming evident. At that time, Component Design Engineering, the group responsible for the design of the company's integrated circuits, (hereafter called *design department*) was organized functionally, with fewer than 100 employees. There were separate groups of architects, logic and circuit designers, layout designers, and groups for product planning, applications, and documentation. More than 80% of these employees had college degrees. Of those who had degrees, at least two-thirds of them had an advanced degree, Master of Science or a Ph.D. The specialist qualification of these employees had shaped the functional nature of the design department organization. This functional division was marked by conflict and friction among the various groups in design department.

Functional specialization among design workers is widespread in Silicon Valley companies. For engineers this creates an elite "caste" of individual "star" performers with predictably high wages. Layout specialists are usually not degree holders, but they do possess special, narrow skills. This specialization of layout designers has promoted high wages for them as a group as well. Individual engineering "stars" and a clannish underclass of layout designers has not encouraged good organizational committment or cooperation in the design of integrated circuits. Most of the companies in the Silicon Valley have been organized in this functional way to recognize and support both types of design specialists. There is a high demand for design people in the electronics industry.

Office Automation in Circuit Design. Computer-assisted design technology was developed for semiconductor industry applications during the 1970s. Efforts to standardize the layout process were very successful. Using computer-assisted design technology, simple, low performance, standardized chips could be laid out as lithographic masks for manufacturing rather easily. Regular, repeatable arrays could be efficiently laid-out with computer-assisted design packages. Computer-assisted design is similiar to the larger family of computer-based graphics technologies, in that the layout lines electronically sketched can be adjusted, moved, tilted, and erased easily and at will. Computer-assisted design packages also permit copying or reproducing a sketch (a line or a whole sub-circuit) to any other location and in any numbers elsewhere on the circuit. Computer-assisted design technology has three primary benefits: It eliminates a digitizing step in producing manufacturing masks, it allows comparing several different solutions to permit a "tighter" layout, in less time, and it permits faster replicating, editing, and correction of layout drawings.

The net effect of computer-assisted design in semi-conductor circuit layout can be seen as a mechanization of tasks without a necessary extension of the designer's individual contribution or commitment to the product. Socio-technical systems management was considered by Zilog management as a way of understanding the needed integration of computer-assisted design technology into the design of more complex products which required more designer commitment than ever before.

In mid-1980 the design department director's key engineering managers attended a socio-technical seminar presented annually by UCLA. These two managers, after returning from the course, reported to their colleagues their interest in pursuing this approach. In the meantime, the director of the design department met with the manager of Zilog's Nampa plant and talked through the values and concepts of socio-technical analysis. Shortly afterwards the design department director decided to use the socio-technical systems analysis to aid in the development of an appropriate organizational design.

After that, events followed in quick succession. The project consultants were selected. A steering committee of Zilog top management people was selected to make certain that the project was supported. A socio-technical systems design team composed of design department employees from each level of the hierarchy and each functional group was assembled by design department management, who selected ten people from among volunteers.

The design team's charter was to complete the socio-technical systems analysis process while meeting periodically with the steering committee to present progress on the project and to identify what support they needed to continue with the process. The socio-technical systems analysis and design process that the team followed was divided into the four steps or phases described previously for the customer service case. They are: (a) A systems scan; (b) A technical analysis and a social analysis; (c) design recommendations; and (d) implementation of design.

The System Scan. The goal of the scan was to clearly state "our organization's purpose," its product, what is outside our organization, what's inside, and its boundaries (both the physical and the technology boundaries). The first things the socio-technical systems design team worked on in the systems scan was the mission statement. They next went through the process of creating a philosophy statement that listed values that are supported throughout the organization. Coming to a consensus regarding the mission and philosophy for the design department was a very difficult task.

The mission statement finally agreed on for the design department was: "To be the leader at developing and supporting integrated circuit products." The company's first product (the "Z-80" microprocessor) had become an industry standard, and the design department wanted to repeat that performance and remain industry leaders. Through the new plant in Idaho, Zilog had a manufacturing organization that was willing to work with and support research and development. The design department analysis team saw that the manufacturing support was in place and hoped the interface with marketing was also there.

The Technical System Analysis. For the engineers on the socio-technical systems team, the technical system analysis was clearly a valuable part of the analysis phase. They quickly agreed that their product was "information" to manufacturing and marketing functions, in place of the more conventionally attractive idea that they produced the circuits themselves. They enjoyed examining the technical variances that had to be controlled. They discovered that there was a world bigger than their own. They had done the scan and had talked about the output of the design department as information for manufacturing. "But," they said "if a product we conceive and develop is different than what is wanted in the market place we aren't going to succeed or survive." How to get that market data, how to do it now became key questions for the team. In the technical system analysis they began analyzing how the current system was operating. The 12 key variances they identified were grouped in three clusters: quality of product, timeliness of product, and other factors such as market acceptance, manufacturability, and number of device features.

In examining the current control of these key variances the socio-technical systems design team also began to realize the importance of acceptance by marketing of whatever they were doing. They became interested in receiving information from outside of their closed system and in understanding what was going on out in the environment. They concluded that design department really did need to work with marketing people.

Benefits of the technical analysis were: (a) clarification about the product of the design system and the key variances, and how those variances were controlled; (b) giving people a common language, the language of the product, to unify their communication and bridge the gaps among their separate languages of specialization; and (c) utilization of specialists to help the design team become knowledgeable about particular elements with which they were unfamiliar.

The Social System Analysis. The social system analysis focused on the logic and circuit design engineers because they were the ones who "lived with" the integrated circuit project for the longest period of time and were most involved with control of the key variances. The socio-technical systems team undertook an analysis of how often the logic and circuit designers talk to other people. The socio-technical systems team was given training in how to develop a questionnaire and how to interview and gather data from all other employees in the design department. They collected data on who talked to whom about what kinds of issues from every employee. All groups interviewed reported a high degree of social contact with peers in their own and other design department units. Most individuals said they received positive support from their own units, but engineers said that layout designers provided poor support. Many respondents mentioned that most people talked about setting schedules, but that few talked about meeting them. Little contact was reported with manufacturing or with corporate staff. Company policy in areas of compensation, performance evaluation, and career planning was criticized. The analysis helped the socio-technical systems team to understand how they could better control key variances (particularly those dealing with timeliness) and how different members of the design department could keep one another informed about those things that were important in improving the results of their work.

Work System Design. The social analysis was completed by the end of summer 1981. It then took 4–5 months before recommendations for change were presented to the design department.

The new design department organization they recommended was one in which layout designers and engineers, with most of the skills required to create a manufacturable integrated circuit design, comprised a product team. The physical move of most of the company, including the design department, to a new location in January 1982, became the unfreezing event during which the new organization for the design department was implemented. With the new product team in the design department organization, most of the drawbacks of the computer-assisted design technology and its use were expected to be eliminated. By this design, the use of the computer-assisted design technology had been adapted for the purposes of the design department system, and not merely for the maximized use-hours of the technology itself. Layout designers and engineers now had a closer relationship, and their adversarial relationship became a thing of the past. Layout designers could understand the whole process better, and they could identify with a single product or family of related products. Additionally, engineers were hired to learn computer-assisted design-based layout techniques, which further improved relations within the product design teams and improved layout effectiveness by further shortening the communication links between circuit design and layout. Design engineers began doing some of their own layout, and in so doing they inspected their own circuit design work. This new organization was based on the realization that the product of the design department was information—information that manufacturing used to manufacture and test the product, and also information that marketing used to sell the product. As the product was being produced, a larger role for technical writers was created to obtain information from the engineers and to train new customers in use of the product. In the marketing departments a separate strategic marketing group was created that performed two functions: (a) helping to create the original concepts for future projects; and (b) helping to introduce those products into the marketplace.

Product teams in design department were developed with the common goals of the responsibility for design from a product concept, and support of an integrated circuit product once designed. The key issue is support, both in terms of marketing and manufacturing. A product team is a group of people with all the necessary skills needed to design that product and who enjoy working with each other. The four teams in the design department have averaged about 10 people each. Team leaders have dealt with work group processes, group member selection, group performance feedback, and group planning. Team leaders negotiate resources on behalf of the team, lead product development, provide direction and integration, manage external boundaries and facilitate activities within the team.

This organizational change effort started in late 1980 and was implemented in early 1982. The current generation of microprocessor projects are multi-year design efforts and have not yet been delivered to market.

Design department management undertook an informal assessment of the new product team organization in June 1983, more than a year after its implementation. Ten of the 40 members comprising the design department's design staff and most of design department's managers met to review their experience with the new product team organization. Fully three-quarters reported that morale was higher because of increased communication and understanding within the teams. Among the factors mentioned as enhancing team morale were member support for one another, increased "ownership" of decisions made by the group (especially by layout designers), and increased awareness and learning about the total design process and its progress. Half of the respondents also expressed mixed feelings and, in some cases, frustration with some part of the effects of the new structure. Some managers complained, for example, that decision making by consensus removed the ability to simply dictate decisions to subordinates, or that it took too long (in an industry where time is always short), even though it led to the "ownership" previously mentioned as a positive aspect. Some engineers questioned the quality of decisions made by a group because the "best" idea may not always be the one accepted. Engineering specialists commented that they had difficulty obtaining layout support from other design teams when they needed it. Meetings that lasted too long or that did not concern them were mentioned as drawbacks by some layout designers. Managers, engineers, and layout designers all commented that the product team structure interfered with close professional relationships among people with similar skills and formal training, and that attempts to deal with this issue had not worked out as well as they wanted.

A telling comment stated by a manager outside design department was that despite the fact that the socio-technical systems analysis and design process took a long time and many man-hours to complete, its usefulness was more essential for the current design projects (such as the 32-bit microprocessor), which required a team structure, than for the more simple projects of the past. He continued that the one-man (or "master-slave") organization that worked for earlier generations of products would simply not be appropriate in the 1980s. He concluded that it was unfair to compare the performance of the new organization with its multi-person, multi-skilled, product orientation to past arrangements over a time during which the integrated circuit products have changed so much.

During the following 2 years since that assessment, many changes in senior management and in the management of design department have occurred. Severe shifts in the company's strategic forecast and planning also have taken place. These shifts in strategy are partly related to these management changes and partly influenced by the dramatic shifts in the semiconductor and microprocessor market. In sum, the engineering personnel have nearly all left the design department, and with them the organizational characteristics described for the department have largely evaporated. The term *engineering teams* and the encouragement of junior engineers to operate the computer-aided design equipment remains, but the interactive teamwork among various engineering functions and specialities is largely gone. The changes attempted in the culture of the design department, surrounded as it was by a largely traditionally managed company, and faced with a volatile labor market and a depressed economy, required the strong support of local management and the well-informed support of corporate management to survive. Because the local management who undertook the changes has departed, and corporate management also has changed several times since 1982, it is little wonder that the design organization has reverted to a more traditional form.

The manufacturing plant in Idaho remains a viable example of a novel and effective work system designed on socio-technical principles. That plant's distance from the corporate offices, and from the volatile labor market of the Silicon Valley surely have been advantages in maintaining that effective but unusual purpose-oriented organization at Nampa.

CONCLUSIONS

Why, given the notable successes of socio-technical systems in manufacturing, has that approach to system design not been applied more to the "office of the future"? The answer lies not in collar color but in the conception of office work as information processing, and the difficulty of seeing the office as a purposive system. Somehow, the leap for many managers from seeing office technology as merely an extension of themselves or of their speciality to seeing themselves and their offices as information processors with a purpose is beyond them. Although the old way was to produce a reliable product in great quantity, new demands for survival surely include adapting appropriately, with committed and informed employees, to changes in the market. Integrated office systems based on socio-technical systems design principles are of obvious benefit in helping a firm adapt, but the low diffusion rate for this innovation suggests that managers still see integrated office technology as a costly "horseless carriage," to reduce clerical or semi-professional staff, or merely of doing old things in new ways. Managers seem not yet aware of the trade-off in costs between maintaining the industrial era control systems and creating the socio-technical systems control systems. The former control systems are expensive and no longer effective, whereas the latter can be proven cost-effective despite their expense.

13 Choices En Route to the Office of Tomorrow

Niels Bjørn-Andersen
Institute of Informatics and Management Accounting
Copenhagen School of Business Administration

Dian Kjærgaard
Department of Business Administration
Odense University, Denmark

ABSTRACT

This chapter presents and illustrates seven fundamental choices that are always made, explicitly or implicitly, when planning, designing, buying, and implementing office technology. The authors challenge users, management, and design teams—especially systems analysts and programmers—to heed the lessons of impact research: The costs and benefits of information technology are determined by technical, social, and political choices, many of which can be made in individual organizations. What is more, technology that supports and develops all white-collar employees—regardless of their status—usually benefits the organization as a whole. Conversely, oppressive technology often is bad for the organization—and for society, as well.

Most of the chapters in this book have reported studies on the impact of information technology (e.g., data processing systems, decision support systems, office automation systems, etc.). The results have shown that there is very little determinism at play. Most changes are contingent upon other factors—both planned and accidental (cf. Bjørn-Andersen, Eason, & Robey, 1986; Kling, 1980b). The ways in which new technology transform white-collar work depend greatly on which technology is chosen, who chooses it, how it is designed and installed, how it is used and what it is used for.

The purpose of this chapter is to summarize what we believe to be the most important lessons of "impact research" in a way that we hope will provoke our readers into carefully considering seven fundamental choices:

1. How broadly one defines the scope of the design and implementation process,

2. How much emphasis is placed on ergonomics and user-friendly interface,

3. Whether the system is to be "people-driven"—in an organic milieu—or "people-driving"—in a mechanical organization,

4. Whether systems are designed to function as replacements for people, monitoring devices or tools,

5. Whether information and education is provided to enable certain or all users to master the technology or only to adapt to it,

6. Whether the process of implementation and design addresses the issue of power imbalance and contributes to equalization or intensification of power differences, and

7. Whether the system is aimed at liberating its users or oppressing them.

These choices are always made—explicitly or implicitly—in the process of designing, selecting, and implementing systems. This is just as true when putting together a system from ready-made standard packages as when the system is to be coded from the bottom up (Eason, 1982; Mumford, 1983; Pava, 1983; Tapscott, 1982).

As no single stakeholder or interest group can make these choices alone, design (including selecting among packages) requires cooperation between rather diverse participants. But experts such as systems analysts and programmers tend to dominate the process and they unfortunately do not seem to be terribly conscious of the consequences of their work or of the many available choices (see especially Mowshowitz 1980). It is hardly surprising, therefore, that so many systems become nightmares—at least for users and management!

We argue that all participants in the design process and users of a system ought to be given the insight and resources necessary to contribute actively in making the right choices. It is these choices that have the most far-reaching implications for how an information system affects organizational structure, job satisfaction, and a host of the other important outcome variables that constitute organizational effectiveness (cf. also Bikson, this volume).

THE SCOPE OF THE DESIGN PROCESS

Most design teams define the design process very narrowly (Bjørn-Andersen & Hedberg, 1977). What they ought to do is to expand the design process to include the issues of ergonomics, job design, and organizational structure. Indeed, creating a good fit between people, organization, and technology may be the most challenging design task today (Bjørn-Andersen, Earl, Holst, & Mumford, 1982; Bjørn-Andersen & Rasmussen, 1980; den Hertog & de Vries, 1977; ETUI, 1981; Kendall, Malecki, Alexander, Wallace, & Wheatley, 1982; Strassman, 1985; Tapscott, 1982).

It is not just a question of being nice to employees. The necessity of a broader scope of design also can be seen in the distribution of costs of an information system. Everyone knows that the relative cost of hardware as a proportion of the total costs of an information system project has plummeted since the 1970s while software costs are rising. What is less well known, but fast becoming evident, is that "org-ware" costs are becoming more and more significant (Bjørn-Andersen 1985; Strassmann 1985). Figure 13.1 suggests how the structure of costs is changing.

Org-ware consists of all the activities and gadgets that are required if an organization is to get an information system that works and that is used. It includes everything from people-oriented activities, such as analysis of user-requirements, education and training, creation of new work procedures and establishment of new social structure, to physical issues like user manuals, furnishings and lighting.

In order to develop good, cost-effective org-ware, it is necessary to understand that an organization is much more than an information processing system. It is a complex social system that cannot be understood by pursuing purely functionalistic paradigms such as the engineering models that characterized nearly every paper in the first International Federation of Information

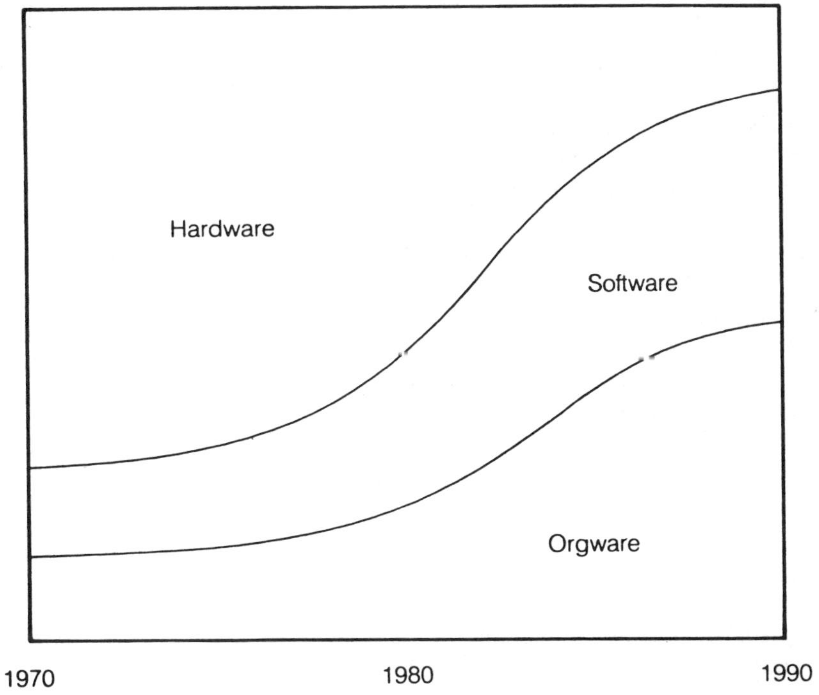

FIGURE 13.1. The distribution of total costs of an information systems project.

Processing (IFIP) world conference on "Human-Computer Interaction" in London in 1984 (a more detailed critique can be seen in Bjørn-Andersen, 1985). An information system must be seen in the context of the larger, "messier" social systems of people who use information.

A useful way of understanding these interrelationships is Leavitt's (1971) model of the mutual dependency of organizational variables: technology, task, structure, and people. These factors exist in a dynamic equilibrium, which we discuss in greater depth later in this chapter (see also Fig. 13.2). A change in any one factor affects all the others. To survive, an organization must seek a new equilibrium every time change occurs. Thus, it is impossible to design a really usable and useful information system without paying attention to organizational design.

Responsibility for this choice should not be left to systems analysts and computer specialists whose education prepares them only to consider the selection, programming, and installation of functioning information systems. Even if they would like to take more responsibility for the human and organizational aspect of their systems, they often are ill prepared. It is vital that management—voluntarily or with prodding—creates support and reward systems to ensure a broad design process. Examples of this kind of support are not unknown (see for instance Bjørn-Andersen, 1984a; Engelbart, 1984; Kling, 1980b; Strassman, 1985), but they are still all too few.

THE CHOICE OF EQUIPMENT AND SOFTWARE

Even the simplest aspects of choosing technology must be seen in relation to people. Ergonomics and user-friendliness are the main issues here. These choices are primarily the responsibility of the design team of computer specialists and users.

Ergonomic standards have been very important in Scandinavia since the early 70s. What could be more obvious than to avoid ruining employees' sight and damaging their backs?

Of course, even the most obvious lessons take time to be learned. It was not until the National Computing Convention in 1981 in Chicago that ergonomic design became an issue in North America.

American and Japanese manufacturers started at that time to catch up with the Scandinavians, who produced ergonomic video display units a good 5 years earlier. However, the lesson had not been completely learned. For instance, some manufacturers marketed their VDUs as ergonomic even though they had non-detachable keyboards! And there were still in 1986 many manufacturers of terminals and microcomputers who break almost every known rule about ergonomics.

Nonetheless, standards for ergonomically well-designed work places (including terminal, furniture, lighting etc.) are reasonably clear and a great variety of affordable equipment is now produced. The interested reader can learn more from Cakir, Hart, and Stewart (1980)—a standard work (also in German, French and Italian)—and HUSAT (1982). Designers nowadays have a real choice and there is very little excuse left for neglecting ergonomics.

In the area of software and user-interface there are a host of other technological choices that are becoming increasingly available and economically feasible. User-friendliness is not just an empty slogan. Helpful instructions and error-messages can make all the difference in how quickly a user learns the system and how much mastery or frustration is experienced (Eason, 1982; Engelbart, 1984; Mehlmann, 1981; Tapscott, 1982; see also Bikson, this volume). Flexible, responsive, and easily integrated applications are a prerequisite for realizing the full potential of information technology and the people who use it.

Existing equipment and systems impose some limitations, to be sure, but these must not be exaggerated. There are tremendous possibilities open to any systems designer who is willing to go beyond an exclusive emphasis on technical performance and who dares to break old habits and ties to manufacturers who haven't kept up with the times.

MECHANICAL AND ORGANIC ORGANIZATIONS

The design of information systems is intimately connected to job and organizational design. Unfortunately, there is an all too widespread prediliction for what we call *people-driving systems.* Such systems require large amounts of formalized information and large numbers of programmed tasks. Thus, they tend to perpetuate and intensify tendencies in an organization to structure itself and behave mechanically. An expansion of Leavitt's model (Fig. 13.2)

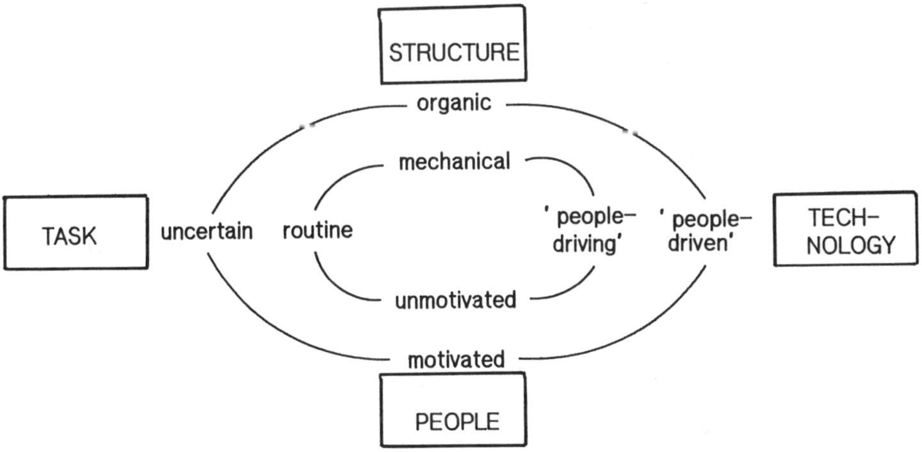

FIGURE 13.2. Mutual dependency of organizational variables.

can help to understand how unfortunate the choice of a people-driving system can be for the organization as a whole.

The inner circle of Fig. 13.2 represents a mechanical organization that also can be called a *Taylorized* organization. Frederich Taylor's *Scientific Management* is based on the principle that work may be carried out more efficiently if it is broken down into very minute sub-tasks, subjected to a detailed analysis so all unnecessary movements are eliminated, and then put together in the most "optimal" way. Planning of work is entirely separated from carrying out work. Workers are just a cog in the machinery that on a day-to-day basis determine work content and intensity (Braverman, 1974; Cooley, 1980; CSS, 1981; see also Kraut, this volume).

People-driving systems are naturally very tempting and rather common in organizations or parts of organizations where task requirements are relatively stable. However, increasingly many organizations find themselves in turbulent environments, where competition and/or high demands for service require highly motivated, skilled, and creative employees, (cf. the outer circle in Fig. 13.2 and Taylor, this volume). As the figure suggests, that is not the kind of employee you get in a Taylorized organization. Organic structures and people-driven technology are strongly correlated with fast-changing environments (Deal & Kennedy, 1982; Galbraith, 1973, 1977; Lawrence & Lorsch, 1967; Mintzberg, 1979).

These findings from empirical research (often identified with the so-called contingency school of organizational theory) often are used normatively to aid survival in today's rapidly changing world. In particular, organizations in dynamic environments are urged to promote movement into the outer circle by using appropriate organization and technology to develop employees. This ought not imply, however, that organizations in more stable environments are best served by maintaining a mechanistic structure and extremely specialized jobs—as on an assembly line or in a word-processing pool.

In the first place, increasingly many organizations are finding it strategically wise to change their environments and thus their structures, tasks, organization, and technology. The motivation for this often is to locate, enter and survive in dynamic and expanding markets. Many organizations also have learned the second important lesson in this connection: People-driving systems are not as economical as they seem, because they create employees who cannot understand the larger process of which their work is a part and who become progressively less motivated and less skilled. This tendency is well documented and criticized in the quality of working life (socio-technical) tradition (e.g., Bjørn-Andersen & Rasmussen, 1980; den Hertog & de Vries, 1977; Mumford, 1983; Pava, 1983; Tapscott, 1982) as well as in more radical traditions (e.g., Braverman, 1974; Cooley, 1980; CSS, 1981; ETUI, 1981), and in very strongly management-oriented traditions (Deal & Kennedy, 1982; Peters & Waterman, 1982).

The attachment to people-driving systems can be surprisingly persistent. In 1980, the Administration Office in West Germany recommended that all government offices create centralized typing pools in order to utilize word-processors optimally, i.e., all day long, as recommended by the leading German manufacturer of word processers. As a result of outside pressure, the Administration Office agreed to carry out a major investigation of the consequences and financed 10 researchers to investigate what the funders hoped would legitimize their decision to create these highly specialized jobs. Halfway through the 2-year study, the researchers had uncovered a very low degree of job satisfaction, very high absenteeism in the centralized typing pools, as well as major problems in communication between authors and typists. However, the Administration Office fired the research team and went on recommending centralized pools—a clear choice!

But there are other alternatives, including new relationships, where secretaries act as collaborators with managers and professionals (Bjørn-Andersen, 1983, 1984a; Mumford, 1983; see also Pava, 1983; Otway & Peltu, 1984). The very least that systems analysts and other professional designers can do is to step back from their love affair with big, beautiful people-driving systems and learn when they are dysfunctional.

In the next two sections we take a closer look at problems with people-driving systems and what it takes to move toward people-driven systems. Knowledge is of course not enough, as the German case illustrates, so the final two sections of this chapter consider political and ideological factors surrounding the design process.

THE FUNCTION OF THE SYSTEM

Results from research on the quality of working life cannot define what the ideal job is. But we can summarize a great deal of knowledge about the dimensions of working life, which can encourage—perhaps even create—active, enthusiastic, committed, creative, and sensible employees.

The most important findings from all types of computer-impact research are classified in Fig. 13.3 (adapted from Bjørn-Andersen & Rasmussen 1980).

With these findings in mind, one can take a critical look at three basic roles an office (or other information) system can play:

1. a device for replacing people,
2. a monitoring device,
3. a tool.

The first two choices are the dominant ones, and many systems designers find themselves more or less forced to create such systems. Whether or not

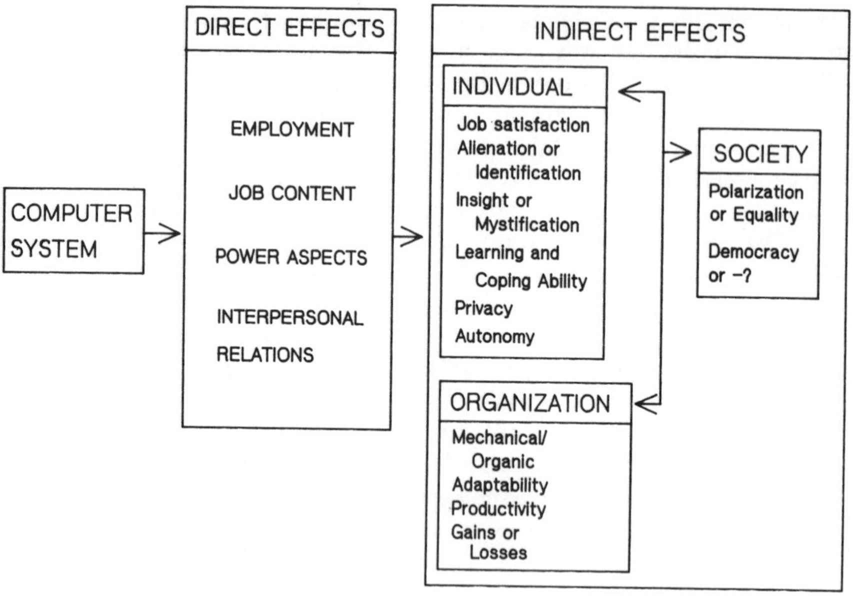

FIGURE 13.3. Quality of working life, organizational effectiveness, and social consequences.

this happens is usually determined by strategic decisions that the design team has to live with. Labor and management are the key actors, and there is a great deal that can be done. For example, monitoring can be prohibited or regulated. Furthermore, collective bargaining can determine how gains from increases in productivity and efficiency are to be shared. Such gains do not always have to be exclusively in the form of higher salaries, wages, and profits, but can also include increased effectiveness, such as better service or a less stressful and more stimulating working environment.

Even when analysts and programmers are operating within relatively narrow confines, they can seek alternatives, especially opportunities to create systems that really work as tools. And designers, users, and managers ought to be more aware of the possible consequences of choosing to replace and monitor people.

Replacing People

Many information systems are deliberately created and installed to automate existing manual procedures and replace workers with computers. The term *office automation* is not accidentally chosen. The attempt to run organizations as economically as possible makes many people redundant. Let us set the social consequences aside until a later section and look at the issues that are easier to see as being the responsibility of a single organization.

In the first place, the fear of being laid off creates tense working environments that can just as easily result in burn-out, apathy, and rigidity as it can result in an elite work-force. Even when the replacement method used is natural attrition, fear and apathy can arise, partly because pressure is sometimes put on people to get out, partly because more socially aware people feel badly about their unemployed colleagues, especially in the younger generation.

In the second place, the substitution of human skill requires the extraction and expropriation of knowledge from individuals to the systems. This process is now spreading to the design of intelligent knowledge-based systems (so-called expert systems and the like). This time around it is not just skilled workers who are affected by classical Taylorism. The new development is that large scale skill substitution can now affect white-collar workers of rather high status, including a variety of experts.

This development can have many justifications in an increasingly complex world, especially when it is necessary to handle large amounts of research results or laws and regulations in order to make a competent decision. But, as Kraut (this volume) points out, the process raises new moral and legal questions about the ownership of information and knowledge.

Such moral and legal issues are far from being the only problems. How and why knowledge is transferred to systems must always be a matter of debate and choice. Again, it is not just a question of being nice to employees who feel that their identities and livelihoods are threatened. Many situations are simply not nearly as predictable and amenable to formalization as some systems designers would like to think. A high degree of automation can be risky if it results in there not being enough people around who are empowered and have the skills to handle unexpected situations.

Monitoring and Control

Management often believes that its interests only can be assured if employees are closely watched. Office automation systems often are promoted on the basis of being ideal for keeping track of every activity, every break, and every mistake. For example, managers in a French insurance company recently decided to do away with their "old-fashioned" keys and substituted them with ID-cards. Every door was equipped with a reader—including the doors to the toilets.

Quality of working life research has, however, confirmed what in our opinion ought to be self-evident. The effects of such monitoring are just as negative as the effects of rigidly people-driving systems. Adults are humiliated and diminished by being watched and controlled at a very detailed level. The negative effects of this situation are well documented among blue-collar workers (see especially Braverman, 1974). Workers with intelligence and energy resist control and sometimes sabotage production. If possible, they seek other kinds of work. The remaining work force becomes indifferent and careless. In the French example, the employees went on strike and stayed there until the system was removed. The insurance company was lucky. Good office work can not be performed by alienated workers, and the effects can undermine the entire organization.

Computers as Tools

If computer systems are to be tools for people, they have to be designed so that employees experience insight into what their work means in the context of the organization (see especially Cooley, 1980; CSS, 1981; den Hertog & de Vries, 1977; Eason, 1982; Engelbart, 1984; ETUI, 1981; Mumford, 1983; Pava, 1983). They ought to be able to use both their education and their experience to affect how they do their work. If they can also influence the content and purpose of their work, so much the better.

For example, expert systems ought to aim at helping the expert to become a better expert. That is, the system should be designed to aid a learning process and should be modifiable so that higher levels of learning and insight are both supported by and incorporated into the system.

More generally, computer systems should create and support a division of labor that stimulates personal development, i.e., which allows employees to follow a job to completion and which guards against unnecessary limitations on their discretion imposed by specialist and management groups. Furthermore, the system must not destroy opportunities for social contact among employees, as a great deal of learning takes places in informal social settings.

Good applications programs are important, but useless unless there is committment on the part of both management and the design team to letting users shape their tools. The movement for end-user programming is a step in the right direction, but seems to be mostly limited to managers and professionals who need interactive decision-support systems, information retrieval programs, and planning tools implemented on personal computers and distributed systems. Let us not forget that secretaries and clerks also can benefit and that not even updating of corporate data bases has to depend on repetitious, deadly dull jobs using people-driving systems.

INFORMATION AND EDUCATION FOR USERS AT ALL LEVELS

It is very common that no one really informs employees about a new system until after it is too late for them to exercise any influence (see e.g., Bikson, this volume). There often are political reasons behind that state of affairs, which we discuss in greater detail later. But even the best of intentions to support employees' influence cannot be fulfilled if employees' education is not good enough.

As a general rule, one should be wary of depending too much on manufacturers' courses. It is a selling point that systems are so easy to use that one only needs a very short time to learn them. Accordingly, courses are typically very short and their aim is usually to teach simple operations and to give an appreciation of their technology. Aiming for an educational level where the employees can modify, evaluate, and even criticize is obviously seldom an objective. There are several quite varied pitfalls with such courses. Sometimes they result in massive frustration, because the participants do not even learn enough to use the equipment in real situations.

In other cases, the courses are quite good at living up to their limited goals. Employees do learn to use their new systems quite quickly and efficiently. But at the same time (Bjørn-Andersen, 1982) no platform is created for future innovations. Indeed, one might even say that resistance to innovation increases. The staff quite rightly consider their qualifications as relevant to the present technology, including all the work procedures that have become enmeshed with the system as a result of consultant recommendations, physical constraints, and long-standing habits. One could of course choose to find ways to rid the organization of such petrified employees, and take in new, similarly poorly trained people for the next technological wave. It is not a choice we would recommend, however.

Quite aside from the objectionable ethical aspects, there are a number of practical drawbacks. Not many organizations can thrive with employees who cannot cope with unusual situations and break-downs. Moreover, it seems a bit absurd to invest in a lot of new technology if employees are not in a position to use all of its relevant possibilities.

In a recent study by the Swedish State Auditors, only 11% of all word-processing and 14% of all personal computer installations in Swedish public administration were found to be cost-effective in 1984 (Administrations Departementet, 1985). The main reason is that the technology is not used to its capacity. For example, many word-processing systems are simply used as slightly more convenient typewriters. The result is sloppier drafts, more numerous revisions, and unrealizable expectations of increased efficiency in typing work. Most of the potential secretarial contributions to the quality of the processed material are lost. Similarly, very little is done to make use of the contributions that could be made by clerical employees who really know what is needed for the development of better filing and retrieving systems. Thus, the technology never gets integrated in a creative or dynamic way into the work process.

The choice to increase end-users' understanding of their systems means that a great deal more must be invested in education—a large part of orgware. There are many possible levels, as summarized in Table 13.1.

The sole responsibility for such education cannot rest with system analysts, programmers, and other design specialists, but they can choose to be a proponent of it and to help carry it out. Management can help with a

TABLE 13.1
Educational Prerequisites for System Design/Implementation

Level of Education	*Allows the User to*
—Appreciation	—Say yes or no
—Operation	—Carry out routine
—Evaluation	—Select
—Understanding	—Modify, choose parameters
—Concrete Mastery	—Program
—Abstract Mastery	—Design

good policy for employee training and development that gives enough time and resources to reach higher goals. At the very least, possibilities for broadening the scope of design and involving employees should be communicated to as many stake holders as possible. For example, there are a number of publications in the Nordic countries that are specifically written for use by employees to design their own job environments (e.g., Arnberg & Bjørn-Andersen, 1984). Analysts and specialists in systems design with a flair for communication can contribute a lot to bringing such publications and the ideas in them into the design process.

Really good user education ought to increase employees' ability to formulate problems, to plan, to program, to deal with complex issues. At the very least, users must have an understanding of technology in relation to their work that allows them to evaluate, criticize, and choose their tools.

In many situations, it is appropriate to educate users to be able to design or change systems themselves (Bikson, this volume; Eason, 1982). Design need not include programming or putting circuit boards together. The most critical parts of design concern choosing what the system is to be used for and how work is to be organized. A certain level of programming skill can be useful, though, and to this end prototyping tools are a marvelous help. But they cannot be fully exploited unless one is interested in making it possible for at least some users to fully master their systems. This choice requires not only that users develop a rather considerable insight into the external properties of technology, but also that both users and technical experts learn how to communicate and cooperate with each other. Skills in group work and cross-disciplinary problem solving are indispensable.

POWER IMBALANCE AND IMPLEMENTATION STRATEGY

It should be quite clear that the interests of an organization are best served through broad design processes to create systems that can be tools for well-educated, committed employees. To achieve this requires employee participation in a democratic design and implementation strategy.

But perhaps our readers will protest that participation must be limited by management's responsibilities and prerogatives. When discussing participation in the design and implementation of three leading systems, a former managing director in the Danish subsidiary of one of the leading United States computer manufacturers remarked, "Oh, yes, we always make sure we have user participation, as we want to have a hostage." We would like to point out, however, that there are many different types of participation that lie between the extremes of taking hostages or letting employees "take over."

The role of information in an organization is varied, complicated, and political. Thus, all white-collar work, and not just managerial work, is a complex social and political activity. Designers of systems must recognize that they are social change agents. Although they usually are hired by managements, they can increase meaningful employee involvement, if they so choose. They must, however, be aware that there are many elements in the systems design process that mitigate against genuine user participation because of the enormous differences in power between system designers and users (see e.g., Bjørn-Andersen, 1984b; Mowshowitz, 1980).

Two kinds of power differences need to be taken into account: direct and indirect. Direct forms of power are the most easily seen, e.g., in the forms of money, time, and mandate. Even though these resources in the final analysis are controlled by management, very large degrees of concession and discretion can be given to employees. Systems analysts and programmers and other highly skilled systems designers often have enough influence to insist on the importance of employee influence—or to make it impossible.

Indirect power is much harder to put a finger on, but there are rich possibilities for systems analysts to do something about it. For instance, take a look at how specialists use language. Do they attempt to confuse, mystify, and intimidate users? Or do they use a language users can understand? Do they attempt to understand users' language and their view of the world? Maslow once said, "to he who has a hammer, the whole world looks like a nail" (in Hoos, 1983, p. 248). Perhaps users do not care to be seen as nails.

Do experts take control over how problems are defined? When problems are defined by systems analysts alone the range of alternatives for users usually gets restricted to a handful of technical options. It is conceivable that there are a large number of alternative, non-technical problem definitions that users would prefer. Systems analysts do not have to define the issue as "which word processor"—so that the issue of "whether or not a word processor" never is taken up with potential users.

Of course, sometimes the problem of indirect power is out of the hands of individual systems designers. The following illustration shows how important it is that the social and legal frameworks for relations between management and labor support democracy at work. However, the example also can show how attitudes, jargon, and definitions of issues can affect balances of power.

In Norway, there is a law on work environment that prohibits the creation of monotonous work. An employer who creates monotonous jobs can be taken to court and fined. There have been at least two cases where a trade union has taken an employer to court for proposing a system that would create unacceptably monotonous jobs. In both cases, the employers decided not to have the issue settled in court. They decided not to implement the offending systems, and the unions dropped the charges.

For a period of 3–4 years, the Danish Employers' Association fought very hard to prevent a similar extension of the Danish Work Environment Law. Their argument was that monotonous work simply cannot be defined. If it cannot be defined, it cannot be negotiated. This was a superbly effective indirect exertion of power. Fortunately—at least we think so—the Danish regulations have now been changed so the issue can be dealt with.

OPPRESSION OR LIBERATION?

When one of us worked on a large study of computers in Danish banks a number of years ago, he asked a teller, "Do you like your new computer system, or would you rather have kept to the old system? She answered, "Oh, I'm not against it, because it's progress, isn't it?" That was a fine selling job on the part of the bank; who can be against progress?

We do not believe that all technical innovations mean progress, if by progress is meant human survival and development. All too many computer systems seem to have nothing to do with that kind of progress and can end up being blind alleys like the Concorde plane—perhaps useful to a tiny elite, but terribly expensive, irritating at best, destructive at worst (for more detailed critique see: Albons, Gunarsson, & Söderborg, 1981; Bjørn-Andersen, 1982; Bjørn-Andersen et al., 1982; Hoos, 1972; Illich, 1973; Kendall et al., 1982; Mander, 1985; Nowotny, 1982.)

It is possible to argue that the organization can afford to make idiots of its blue-collar workers, but what of the long run? And can society afford it? With the spread of exaggerated control into white-collar work, the issues become, if possible, even more critical: The evidence is overwhelming that organizational effectiveness requires employees who are competent adults and who are treated as such.

It also is important to be aware of the fact that many of the skills that people develop at work also are important to maintaining democracy. Well-educated and demanding work forces are a treasure. Information systems should not erode that tremendous asset by impoverishing the quality of working life.

Lay-offs create considerable social disruption. A large proportion of the redundant will have to be retrained and redeployed, and a certain number of them will be doomed to poverty and a sense of being useless. The concentration of productivity gains in very few hands can come to polarize the population into a work force versus recipients of support. Polarization of the work force itself also is a threat, particularly for women and for less well-educated men and minority groups, who are limited to the most alienating and deadening jobs, whereas well-educated white men enjoy the fruits of technology. And polarization between information-rich and information-poor countries is already so marked, that UNESCO in 1984 called for a new "Information Order" to bridge the widening gap.

Everyone who has the slightest possibility of affecting the first six choices ought to think seriously about who is really benefiting from so-called technical achievements, and who is paying the price: Capital or labor? The well-educated or those without education? The established generations or their children? Experts or nonexperts? Men or women? Developed or developing countries? Is the information society for richer or poorer?

Whether we are conscious of it, all of us make ethical choices all the time. Systems designers can choose to work on systems and implementation strategies that reduce inequalities and even actively oppose oppressive systems. It is true that a more equal world cannot be created by systems designers alone, but the possibilities are greater than many would believe. The choice is up to us.

References

Adler, P. (1984). *New Technologies, New Skills.* (Working Paper 9–784–076). Harvard Business School: Cambridge, MA.

Administrations Department. (1985). [Office technology in public administration: A learning process]. *"KT-90": Kontonteknologi i statenen laereproces.*

The Aetna affiliated companies: a brief historical sketch. (1923, January). *The Aetna-izer, 54.*

Aetna Life Insurance Company. (n.d.). *Instructions to division heads.* Typewritten circular, Aetna Life and Casualty Company archives, Hartford, CT.

Aetna Life Insurance Company. (1908). *Rules and regulations of the home office: Effective July 1, 1908* (Circular No. 509). Pamphlet, Aetna Life and Casualty Company archives.

Aetna Life Insurance Company, Employment Registration Bureau. (1929–1930) [Records, December 29, 1929–November 27, 1930]. File, Aetna Life and Casualty Company archives.

Aetna Life Insurance Company, Aetna Accident and Liability Company, Automobile Insurance Company of Hartford. (1915). *Rules and regulations of the home office.* Pamphlet, Aetna Life and Casualty Company archives.

AFL-CIO (1983). *The future of work. A report by the AFL-CIO Committee on the evolution of work.* Washington, DC: Author.

Albons, D., Gunnarsson, E., & Söderberg, I. (1981). [For Richer or poorer: A report on computerisation in the US]. *På vinst och Förlust—rapport om datorisering i USA.* Stockholm: TCO (Tjanestmannens Central Organisation—Central Organisation for Civil Servants).

American Telephone & Telegraph. (1982). *The structure of the work-at-home market: Job/volunteer/school.* (Unpublished manuscript). Basking Ridge, NJ: Author.

An American miracle that works. (1982, November). *Productivity, 3,* 1–5.

Appelbaum, E. (1984). *The impact of technology on skill requirements and occupational structure in the insurance industry, 1960–1990.* (Unpublished paper). Philadelphia: Temple University, Department of Economics.

Appelbaum, E. (1985). *Technology and the redesign of work in the insurance industry.* Stanford: Stanford University Institute for Research on Educational Finance and Governance.

Applegath, J. (1981). *Working free.* Washington, DC: World Future Society Press.

Arnberg, B., & Bjørn-Andersen, N. (1984). [Working with a VDU]. *Dit arbejde ved terminalen. Copenhagen: Samfundslitteratur.*

Aronson, S. (1977). Bell's electrical toy: What's the use? The sociology of early telephone usage. In I. de Sola Pool, (Ed.). *The social impact of the telephone.* (pp. 15–39). Cambridge, MA: MIT Press.

Attewell, P., & Rule, J. (1984). Computing and organizations: What we know and what we don't know. *Communications of the ACM, 27,* 1184–1192.

Audibert, A. (1984, September). *Office automation at SERETE.* Paper presented at EUROJOB 84, SERETE, 86 rue Regnault, 75640 Paris.

Bair, J. H. (1978). *Productivity assessment of office information systems technology.* Paper presented at the IEEE Conference on Trends and Applications 1978: Distributed Processing, New York.

Bair, J. H. (1979a). Communication in the office of the future: Where the real payoff may be. *Business Communication Review, 9,* 3–11.

Bair, J. H. (1979b). *Productivity assessment of office automation systems.* Menlo Park, CA: SRI International Reports.

Bair, J. H. (1980, Spring). Avoiding working non-solutions to office communications systems design. *Com Con,* 99–103.

Baker, E. (1973). *Technology and women's work.* New York: Columbia University Press.

Baran, B. (1984). *Office automation and women's work: The technological transformation of the insurance industry.* (Unpublished paper). Berkeley: University of California, Department of City and Regional Planning.

Baran, B. (1985, February–March). *The technological transformation of white collar work: A case study of the insurance industry.* Paper prepared for the Panel on Technology and Women's Employment Workshop, National Academy of Sciences: Washington, DC.

Baran, B., & Teegarden, S. (1984). *Women's labor in the office of the future.* (Unpublished manuscript). Berkeley: University of California, Department of City and Regional Planning.

Bayman, P., & Mayer, R. E. (1984). Instructional manipulation of users' mental models for electronic calculators. *International Journal of Man-Machine Studies, 20,* 189–199.

Becker, F., & McClintock, C. (1981, March). *Mixed blessings: The office at home.* Paper presented at the National Telecommunications Conference, Houston.

Behr, M., & Lazar, W. (1983). *Women working home: The home-based business guide and directory.* Summit, NJ: Women Working Home Press.

Bell Communications Research News. (1985, March 8). *Integrated services digital network.* Livingston, NJ: Bell Communications Research.

Benoit, C., Cossette, A., & Cardillo, P. (1984). *L'incidence de la machine a traitement de textes sur l'emploi et le travail.* Montreal, Canada: Ministère du Travail.

Berman, P., & McLaughlin, M. (1978). *Federal programs supporting educational change: Factors affecting implementation and continuation. (Report No. R-1589/7-HEW).* Santa Monica, CA: Rand Corp.

Biderman, A. D., & Drury, T. F. (Eds.). (1976). *Measuring work quality for social reporting.* New York: Halsted Press.

Bielby, W., & Baron, J. (1985). A women's place is with other women: Sex segregation in the workplace. In B. Reskin (Ed.), *Sex segregation in the workplace.* (pp. 27–55). Washington, DC: National Academy Press.

Bikson, T. K. (1980). *Getting it together: Gerontological research and the real world (Report No. P-6447).* Santa Monica, CA: The Rand Corp.

Bikson, T. K., & Eveland, J. D. (1985). *New technology in the office: Planning for people.* Scarsdale, NY: Work In America Institute.

Bikson, T. K., & Gutek, B. A. (1983). *Advanced office systems: An empirical look at utilization and satisfaction* (Report No. N-1970-NSF). Santa Monica, CA: The Rand Corp.

Bikson, T. K., & Gutek, B. A. (1983b). *Training in automated offices: An empirical study of design and methods.* Paper presented at IFAC/IFIP Training for Tomorrow Conference. Leiden, The Netherlands.

Bikson, T. K., Gutek, B. A., & Mankin, D. A. (1981). *Implementation of information technology in office settings: Review of relevant literature. (Report No. P-6697).* Santa Monica, CA: The Rand Corp.

Bikson, T. K., Gutek, B. A., & Mankin, D. A. (1985). *Implementing computerized procedures in office settings: Influences and outcomes (Report No. R-3077-NSF).* Santa Monica, CA: The Rand Corp.

Bikson, T. K., Quint, B., & Johnson, L. L. (1984). *Scientific and technical information transfer: Issues and options (Report No. N-2131-NSF).* Santa Monica, CA: The Rand Corp.

Bikson, T. K., Stansz, C., & Mankin, D. A. (1985). *Computer-mediated work: Organizational and individual consequences in one corporate headquarters (Report No. R-3313-OTA).* Santa Monica, CA: The Rand Corp.

Bills, M. A. (1927). An application of the individual bonus plan to home office clerical work. In *Proceeding of the 1927 annual conference of the Life Office Management Association* (pp. 153–159, 170–172). New York: Life Office Management Association.

Bills, M. A. (1928). Time study as a basis for measuring office output. In M. A. Bills, A. C. Farrell, & H. B. Hill (Eds.), *Measuring office output.* (Office executives' series, 32, pp. 2–6). New York: American Management Association.

Bills, M. A. (1929, November). *Measuring, standardizing and compensating for office operations.* Paper presented at the Office Conference, American Management Association Autumn Convention, Detroit. (Office management series, 44). New York.

Bills, M.A., Wallace, C., Donaldson, A. S., & Lies, B. E. (Eds.). (1926). *Measuring office output.* (Office executives' series). New York: American Management Association.

Bjørn-Andersen, N. (1982, June). *Work qualifications and educational requirements en route for the information society.* Paper presented to workshop at the European Centre for the Development of Vocational Training (CEDEFOP), Berlin.

Bjørn-Andersen, N. (1983). The changing roles of secretaries and clerks. In H. Otway & M. Peltu (Eds.), *New office technology: Human and organizational aspects* (pp. 120–137). Norwood, NJ: Ablex.

Bjørn-Andersen, N. (1984a). Expected effects of office automation. In *Proceedings of the International Conference on Trends in Office Automation in Local Authorities, September 1984* (pp. 145–158). Copenhagen: Kommunedata/IULA (International Union of Local Authorities).

Bjørn-Andersen, N. (1984b). Information technology and power change in organizations—Prospects of technology agreements and technology assessments. In K. W. Grewlich & F. H. Pedersen (Eds.), *Power and participation in an information society* (pp. 183–202). Commission of the European Communities: FAST (Forecasting and Assessment in the field of Science and Technology) in cooperation with Ministry of Industry, National Agency of Technology, Council of Technology in Denmark. (EUR 8545 EN).

Bjørn-Andersen, N. (1985). Are "human factors" human?: In D. Murray & N. Bevan (Eds.) *Man-machine integration: The coming of age* (pp. 16–23). London: Pergamon.

Bjørn-Andersen, N., Earl, M., Holst, O., & Mumford, E. (Eds.). (1982). *Information society—For richer, for poorer.* Information Research and Resource Reports Vol. 2. Amsterdam: North-Holland.

Bjørn-Andersen, N. Eason, K., & Robey, D. (1986). *Managing computer impact. An international study of organizations and management.* Norwood, NJ: Ablex.

Bjørn-Anderson, N., & Hedberg, B. L. T. (1977). Designing information systems in an organizational perspective. In P. C. Nystrom & W. H. Starbuck (Eds.), *Studies in the management sciences. Prescriptive models of organizations* Vol. 5, (pp. 125–142). Amsterdam: North-Holland.

Bjørn-Andersen, N., & Rasmussen, L. B. (1980). Sociological implications of computer systems. In H. T. Smith & T. R. G. Green (Eds.), *Human interaction with computers.* (pp. 97–123). London: Academic Press.

Blomberg, J. (1983, November). *The social organization of the office: Implications for successful placement of office information systems.* Paper presented at the 82nd Annual Meeting of the American Anthropological Association, Chicago, IL.

Brainard, M. B. (1931). [Memorandum to all officers and division heads of the Aetna Life affiliated companies, April 9]. Unpublished manuscript, Aetna Life Insurance Company, Circular Letters, Series C, Vol. 7, p. 140, Aetna Life and Casualty Company archives.

Braverman, H. (1974). *Labor and monopoly capital: The degradation of work in the twentieth century.* New York: Monthly Review Press.

Brewer, G. D. (1980). On the theory and practice of innovation. *Technology in Society, 2,* 337–363.

Burris, V., & Wharton, A. (1982). Sex segregation in the U.S. labor force. *The Review of Radical Political Economics, 14,* 43–56.

Busterer, K. (1979). *Know-how on the job.* Boulder, CO: Westview Press.

Cadbury, E., Matheson, M., & Shann, G. (1906). *Women's work and wages.* London: T. Fisher Unwin.

Cakir, A., Hart, D., & Stewart, T. F. M. (1980). *Visual display terminals.* London: Wiley.

Cammann, C., Fichman, M., Jenkins, G. D., & Klesh, J. R. (1978). *The Michigan Organizational Assessment Questionnaire, Michigan Organizational Assessment Package.* Ann Arbor, MI: Survey Research Center, Institute for Social Research, University of Michigan.

Caporael, L. (1985). Computers, prophecy, and experience: A historical perspective. *Journal of Social Issues, 40,* 15–31.

Card, S., Moran, T., & Newell, A. (1981). *The psychology of human-computer interaction.* Hillsdale, NJ: Lawrence Erlbaum Associates.

Carey, M. (1981). Occupational employment growth through 1990. *Monthly Labor Review, 104,* 42–55.

Carey, M. (1982). Occupational employment growth through 1990. In U.S. Department of Labor, Bureau of Labor Statistics, *Economic projections to 1990.* Washington, DC: U.S. Government Printing Office.

Casparis, J. (1983). Switzerland: Demand for labor from the Mediterranean. Paris: 2nd Colloquium, Political Economy of Southern Europe.

Chabrow, E. (1985, April). Telecommuting: Managing the remote workplace. *Information Week,* 27–35.

Chamot, D. (1985). Labor and technological chance in Sweden. In *Working life in Sweden,* (No. 29). New York: Swedish Information Service.

Chamot, D., & Zalusky, J. (1985). Use and misuse of workstations at home. In *Office workstations in the home* (pp. 76–84). Washington, DC: National Academy Press.

Cheney, P. H., & Dickson, G. W. (1982). Organizational characteristics and information systems: An exploratory investigation. *Academy of Management Journal, 25,* 170–184.

Cherry, L., & Vesterman, W. (1979). *Writing tools: The STYLE and DICTION programs.* AT&T Bell Laboratories, Technical Memorandum.

Christensen, K. (1984). *Impacts of home-based work on women and their families.* Washington, DC: Department of Health & Human Services.

Cockroft, D. (1980). New office technology and employment. *International Labor Review, 119,* 689–704.

Cooley, M. (1980). *Architect or bee?.* Slough, England: CAITS.

Council for Science and Society. (1981). *New technology: Society, employment and skill.* London: Blackrose Press.

Cuff, R. N. (1980). On casual users. *International Journal of Man-Machine Studies, 12,* 163–187.

Culnan, M. J., & Bair, J. H. (1983). Human communication needs and organizational productivity: The potential impact of office automation. *Journal of the American Society for Information Science, 34,* 215–221.

Cummings, L. (1977). *The rationalization and automation of clerical work.* Unpublished master's thesis, Brooklyn College.

Dainoff, M. (1982). Occupational stress factors in visual display terminals (VDT) operation: A review of empirical research. *Behavior and Information Technology, 1,* 141–176.

Danziger, J. N., & Dutton, W. H. (1977). Computers as an innovation in American local governments. *Communications of the ACM, 20,* 945–956.

Davies, M. (1983). *Woman's place is at the typewriter: Office work and office workers 1870–1930.* Philadelphia: Temple University Press.

Davis, L. E. (1981). Organization design. In G. Salvendy (Ed.), *Handbook of industrial engineering* (pp. 2.1.1–2.1.29). New York: Wiley.

Davis, L. E., & Taylor, J. C. (1976). Technology, organization, and job structure. In R. Dubin (Ed.), *Handbook of work, organization, and society* (pp. 379–419). Chicago: Rand McNally.

Deal, T. E., & Kennedy, A. A. (1982). *Corporate cultures. The rites and rituals of corporate life.* Reading, MA: Addison-Wesley.

den Hertog, J. F., & de Vries, H. J. J. (1977). *Breaking the deadlock: Search for new strategies for Q.W.L.* The Netherlands: Philips' Gloeilampenfabrieken, Industrial Psychological.

de Kadt, Maarten. (1979). Insurance: A clerical work factory. In Andrew Zimbalist (Ed.), *Case studies on the labor process* (pp. 242–256). New York: Monthly Review Press.

Delgado, A. (1979). *The enormous file: A social history of the office.* London: John Murray.

Dixon, C. (1965). [Memorandum to Elena N. Patrini, January 28]. Unpublished manuscript, Topical file "Employees," Aetna Life and Casualty Company archives.

Driscoll, J. (1980, March). *Office automation: The dynamics of a technological boondoggle.* Paper presented at the International Office Automation Symposium, Stanford University.

Driscoll, J. W. (1978). People and the automated office. *Datamation, 25*(12), 106–112.

Drucker, P. (1959). *Technology, management, society.* New York: Harper & Row.

Dubnoff, S., & Kraft, P. (1984, March). *Gender stratification in computer programming.* Paper presented to the Eastern Sociological Association, Boston, MA.

Eason, K. D. (1982). The process of introducing information technology. *Behaviour and Information Technology, 1,* 197–213.

Edwards, P., & Edwards, S. (1985). *Working from home.* Los Angeles: Jeremy P. Tarcher.

Electronic Services Unlimited. (1984). *Telework: A multi-client study.* New York: Author.

Emery, F. E., & Emery, M. (1978). Searching. In J. W. Sutherland (Ed.), *Management handbook for public administrators* (pp. 245–263). New York: Van Nostrand Reinhold.

Emery, F. E., Foster, M., & Wollard, W. (1978). Analytical model for socio-technical systems, 1967. In F. E. Emery (Ed.), *The emergence of a new paradigm of work* (pp. 95–106). Canberra: Centre for Continuing Education, Australian National University.

Emery, F. E., & Trist, E. L. (1965). The causal texture of organizational environments. *Human Relations, 18,* 21–32.

Engelbart, D. (1984, September). From automating the office will come augmenting the organization. In *Proceedings of the International Conference on Trends in Office Automation in Local Authorities, September 1984* (pp. 7–18). Copenhagen: Kommunedata, (International Union of Local Authorities).

Engelbart, D. C., Watson, R. W., & Norton, J. C. (1973, June). The augmented knowledge workshop. *American Federation of Information Processing Societies Conference Proceedings,* (Vol. 42, pp. 9–21). National Computer Conference.

European Trade Union Institute. (1981). *Redesigning jobs: Western European experiences.* Brussels.

Eveland, J. D., & Rogers, E. M. (1980). Measuring the innovation process in organizations. *Journal of Technology Transfer, 4,* 1–13.

Feldberg, R., & Glenn, E. N. (1979). Male and female: Job vs. gender models in the sociology of work. *Social Problems, 25,* 52–64.

Feldberg, R. L. and Glenn, E. N. (1983). Technology and work degradation: Effects of office automation on women clerical workers. In Joan Rothschild (Ed.), *Machina ex dea: Feminist perspectives on technology* (pp. 59–78). New York: Pergamon Press.

Festinger, L., Schachter, S., & Back, K. (1950). *Social pressures in informal groups: A study of human factors in housing.* New York: Harper & Row.

Forester, T. (Ed.). (1981). *The microelectronics revolution.* Cambridge, MA: MIT Press.

Form, W., & McMillan, D. (1983). Men, women, and machines. *Sociology of Work and Occupations, 10,* 147–178.

Fuller, Walter D. (1919, June). Application of scientific principles to office management. *Bulletin of the Taylor Society, 4,* 8–28.

Galbraith, J. (1973). *Designing complex organisations.* Reading, MA: Addison-Wesley.

Galbraith, J. (1977). *Organisation design.* Reading, MA: Addison-Wesley.

Geisler, G. (1985). Blue Cross/Blue Shield of South Carolina: Program for clerical workers. In *Office workstations in the home* (pp. 16–23). Washington, DC: National Academy Press.

Guiliano, V. (1981a). Teleworking: A prospectus. *Telephony, 200,* 67–75.

Guiliano, V. (1981b). Teleworking: Future shock. *Telephony, 200,* 56–62.

Giuliano, V. (1982). The mechanization of office work. *Scientific American, 247,* 148–164.

Glenn, E. N., & Feldberg, R. L. (1979). Degraded and deskilled: The proletarianization of clerical work. *Social Problems, 26,* 524–538.

Goffman, E. (1961). *Asylums.* Garden City, NY: Anchor.

Gottmann, J. (1977). Megalopolis and antipolis: The telephone and the structure of the city. In Pool, I. (Ed.), *The social impact of the telephone* (pp. 303–317). Cambridge, MA: MIT Press.

Gould, J. (1981). Composing letters with computer-based text editors. *Human Factors, 23,* 593–606.

Grandjean, B. D., & Taylor, P. A. (1980). Job satisfaction among clerical workers: Status panic or the opportunity structure of office work? *Sociology of Work and Occupations, 7,* 33–53.

Greenbaum, J. (1979). *In the name of efficiency: A study of change in data processing work.* Philadelphia: Temple University Press.

Gregory, J., & Nussbaum, K. (1982). Race against time: Automation of the office. *Office: Technology and People, 1,* 197–236.

Grief, I. (1985, February). Computer-supported group work: What are the issues? *Office Automation Conference Digest* (pp. 73–76). Atlanta, GA: American Federation of Information Processing Societies.

Gustavson, P. W., & Taylor, J. C. (1984). Sociotechnical design and new forms of work organization: Integrated circuit fabrication. In F. Butera & J. Thurman (Eds.), *Automation and work design* (pp. 697–713). Amsterdam: North Holland.

Gutek, B. (in press). Women in clerical work. In A. Armstrong & S. Harkness (Eds.), *Women working* (2nd ed.). Palo Alto: Mayfield.

Hacker, S. L. (1979). Sex stratification, technology and organizational change: A longitudinal case study of AT&T. *Social Problems, 26,* 539–557.

Hackman, J. R., & Oldham, G. R. (1975). The development of the job diagnostic survey. *Journal of Applied Psychology, 60,* 159–170.

Hackman, J. R., & Oldham, G. R. (1980). *Work Redesign.* Reading, MA: Addison-Wesley.

Hackman, J. R., Pearce, J. L., & Wolfe, J. C. (1978). Effects of change in job characteristics on work attitudes and behaviors: A naturally occurring quasi-experiment. *Organizational Behavior and Human Performance, 21,* 289–304.

Hammer, M. (1984, May). *Telecommunications—The glue of office automation.* Seminar presented to the Los Angeles Office Automation Council, Marina Del Rey, CA.

Harkness, R. C. (1977). *Technology assessment of telecommunications-transportation interactions.* Menlo Park, CA: Stanford Research Institute.

Hathaway, H. K. (1923, April). Premium and bonus plans. *Bulletin of the Taylor Society, 8*(2), 59–65.

Helander, M., & Rupp, B. (1984). An overview of standards and guidelines for visual display terminals. *Applied Ergonomics, 3,* 185–195.

Helfgott, R. B. (1966). EDP and the office work force. *Industrial and Labor Relations Review, 19,* 503–516.

Henderson, D. A., Jr. (1983, December). *Trillium: A user interface design environment.* Paper presented at CHI '83 Conference on Human Factors in Computing Systems, Videotape Session: Noteworthy Interface Designs 1, Boston, MA.

Hewes, J. J. (1981). *Worksteads: Living and working in the same place.* New York: Doubleday.

Hippel, E. von. (1976). The dominant role of users in the scientific instrument innovation process. *Research Policy, 5,* 212–239.

Hippel, E. von. (1983). *Novel product concepts from lead users: Segmenting users by experience.* (Working paper no. 1476–83). Cambridge, MA: Massachusetts Institute for Technology, Sloan School of Management.

Honeywell Corporation. (1983). *Office automation and the workplace.* Minneapolis, MN: Author.

Hooker, Richard. (1956). *Aetna Life Insurance Company: its first hundred years.* Hartford: Aetna Life Insurance Company.

Hoos, I. R. (1961). *Automation in the office.* Washington, DC: Public Affairs Press.

Hoos, I. (1983). *Systems analysis in public policy: A critique.* Berkeley: University of California Press.

Hoppock, R. (1935). *Job satisfaction.* New York: Harper & Row.

Hunt, T., & Hunt, H. A. (1985a). *A compendium of data sources for studying the employment effects of technology.* Washington, DC: National Academy Press.

Hunt, T., & Hunt, H. A. (1985b, April). *Clerical employment and technological change: A review of recent trends and projections.* Paper prepared for the Panel on Technology and Women' Employment, National Academy of Sciences, Washington, DC.

HUSAT Research Group—Loughborough University of Technology, (1982). *Health hazards of VDUs.* London: Wiley.

Iacono, S., & Kling, R. (1984). Office routine: The automated pink collar. *IEEE Spectrum, 21,* 73–76.

Iacono, S., & Kling, R. (1985). *Why organizations adopt new office technologies* (Working paper). Irvine: University of California, Irvine, Public Policy Research Organization.

If home is where the worker is. (1982, May). *Business Week,* 66.

Illich, I. (1973). *Tools for conviviality.* London: Harper & Row.

The instant offshore office. (1980, July). *Business Week.*

Ives, H. E. (1930) Two way television. *Bell Laboratory Record, 8,* 399–404.

James, L. R., & Jones, A. P. (1976). Organizational structure: A review of structural dimensions and their conceptual relationships with individual attitudes and behavior. *Organizational Behavior and Human Performance, 16,* 74–113.

Johnson, B., Eveland, J., Taylor, J., Rice, R., Cline, T., Smith, D., & Browning, L. (1985). *Innovation in office systems implementation* (Report No. 811079). Washington, DC: National Science Foundation.

Johnson, B. M., & Taylor, J. C. (1985). Information processing and career opportunities for women. In A. Olerup, L. Schneider, & E. Monod (Eds.), *Women, work and computerization.* Amsterdam: North Holland.

Johnson, L. (1982). *The seam allowance: Industrial homework in Canada.* Toronto: The Women's Press.

Kaiser, K., & Srinivasan, A. (1982). User-analyst differences: An empirical investigation of attitudes related to systems development. *Academy of Management Journal, 25,* 630–646.

Kanter, R. M. (1977). *Men and women of the corporation.* New York: Basic Books.

Katz, E., Levin, M. L., & Hamilton, H. (1963). Traditions of research on the diffusion of innovation. *American Sociological Review, 28,* 237–252.

Kendall, P. M. M., Malecki, A. M. J., Alexander, W., Wallace, A. S., & Wheatley, T. F. (1982). *The impact of chip technology on the conditions and quality of work—Worldwide search and examination of evidence & influential opinion* (Report # 1144). Den Haag: Ministerie van Sociale Zaken in Werkgelegenheid.

Kiesler, S. (1984). Computer mediation of communication. *American Psychologist, 39,* 1123–1134.

Kiger, J. I. (1984). The depth/breadth trade-off in the design of menu-driven user interfaces. *International Journal of Man-Machine Studies, 20,* 201–213.

Kimberly, J. R., & Evanisko, M. J. (1981). Organizational innovation: The influence of individual, organizational, and contextual factors on hospital adoption of technological and administrative innovation. *Academy of Management Journal, 24,* 689–713.

King, W. R., & Rodreguez, J. I. (1981). Participative design of strategic decision support systems: An empirical assessment. *Management Science, 27,* 717–726.

Kling, R. (1980a). *The impacts of computing on the job characteristics of managers, data analysts and clerks* (Working paper). Irvine: University of California, Public Policy Research Organization.

Kling, R. (1980b). Social analysis of computing: Theoretical orientations in recent empirical research. *Computing Surveys, 12,* 61–110.

Kling, R. (1983). Social goals in systems planning and development. In H. J. Otway & M. Peltu (Eds.), *New office technology: Human and organizational aspects* (pp. 221–237). Norwood, NJ: Ablex.

Kling, R. (1984). Assimilating social values in the design of computer systems. *Telecommunications Policy, 8,* 127–148.

Kling, R., & Iacono, S. (1984). Computing as an occasion for social control. *Journal of Social Issues, 40,* 77–96.

Kling, R., & Iacono, S. (in press-a). Computerization as the product of social movements. In R. Gordon (Ed.), *Microelectronics in transition.* Norwood, NJ: Ablex.

Kling, R., & Iacono, S. (in press-b). Desktop computerization and transformations of work in future offices. In R. Gordon (Ed.), *Automation, work, and employment.* Berkeley, CA: University of California Press.

Kling, R., & Scacchi, W. (1982). The web of computing: Computer technology as social organization. *Advances in Computers, 21,* 1–90.

Kraemer, K. (1982). Telecommunications-transportation substitution and energy productivity: A re-examination. *Telecommunications Policy, 6,* 39–59.

Kraft, P. (1977). *Programmers and managers.* New York: Springer-Verlag.

Kraft, P., & Dubnoff, S. (in press). Division of labor, task fragmentation, and hierarchy in computer software work. *Industrial Organization.*

Kraut, R. E. (in press). Telework as a work-style innovation. *Information and Behavior, 2.*

Kraut, R. E., & Grambsch, P. (1985, June). *Prophecy by analogy: Potential motivations for and consequences of electronic homework.* Paper presented at the Computer and Society Conference, University of Rochester, Rochester, NY.

Lawrence, P. R., & Lorsch, J. W. (1967). *Organization and environment.* Cambridge, MA: Harvard University Press.

Leavitt, H. J. (1971). Change in industry: Structural, technical and human approaches. In V. H. Vroom & E. L. Deci (Eds.), *Management and motivation.* Middlesex: Penguin Books.

Leffingwell, W. H. (1917). *Scientific office management.* Chicago: A. W. Shaw.

Leffingwell, W. H. (1925). *Office management: Principles and practice.* Chicago & New York: A. W. Shaw.

Leontieff, W., & Duchin, F. (1984, April). *The impacts of automation on employment, 1963–2000.* Unpublished final report, New York: Institute for Economic Analysis.

Lepper, M. (1985). Microcomputers in education: Motivational and social issues. *American Psychologist, 40,* 1–18.

Levine, M. F. (1983a). Self-developed quality of working life measures. *Journal of Occupational Behavior, 4,* 35–47.

Levine, M. F. (1983b). Self-developed quality of working life measures: Act II. *Proceedings of the National Academy of Management Annual Meeting,* Dallas.

Levine, M. F., Taylor, J. C., Davis, L. E. (1984). Defining quality of working life. *Human Relations, 37,* 81–104.

The Life Aetna-izer. (1931, November).

Life Office Management Association. (1928). *Proceeding of the 1928 annual conference of the Life Office Management Association.* New York: Life Office Management Association.

Locke, E. (1976). The nature and causes of job satisfaction. In M. Dunnette (Ed.), *Handbook of industrial and organizational psychology* (pp. 1297–1349). Chicago: Rand McNally College Publishing.

Lucas, H. (1975). *Why information systems fail.* New York: Columbia University Press.

Lucas, H. (1978). Unsuccessful implementation: The case of a computer-based order entry system. *Decision Sciences, 9,* 68–79.

Maddox, B. (1977). Women and the switchboard. In I. Pool (Ed.), *The social impact of the telephone* (pp. 262–280). Cambridge, MA: MIT Press.

Makower, J. (1981). *Office hazards: How your job can make you sick.* Washington, DC: Tilden.

Mander, J. (1985). Six grave doubts about computers. *Whole Earth Review, 44,* 10–20.

Mankin, D., Bikson, T. K., & Gutek, B. (1983). Factors in the successful implementation of computer-based office information systems. *Journal of Organizational Behavior Management, 6,* 1–20.

Mann, F. C., & Williams, L. K. (1960). Observations of the dynamics of a change to electronic data processing equipment. *Administrative Science Quarterly, 5,* 217–256.

Market size is underestimated. (1977). *Datamation, 23*(9), 161.

Markus, M. L. (1984). *Systems in organizations: Bugs and features.* Boston: Pitman.

Matteis, R. J. (1979). The new back office focuses on customer service. *Harvard Business Review, 57,* 146–159.

Matthaei, J. A. (1982). *An economic history of women in America: Women's work, the sexual division of labor, and the development of capitalism.* New York: Schocken Books.

McClintock, C. (1981, March). *Working alone together: Managing telecommuting.* Paper presented at National Telecommunications Conference, Houston.

McDavid, M. (1985). U.S. Army: Prototype program for professionals. In *Office Workstations in the Home* (pp. 21–32). Washington, DC: National Academy Press.

McDermott, R., & Roth, D. (1978). The social organization of behavior: Interactional approaches. *Annual Review of Anthropology, 7,* 321–345.

Mehlmann, M. (1981). *When people use computers. An approach to developing an interface.* Englewood Cliffs, NJ: Prentice-Hall.

Melville, H. (1959). Bartleby. In *Four short novels.* New York: Bantam Books. (Originally published in 1856).

Miller, D. (1983, October). *The IBM teleworking experiment. Paper presented at the National Executive Forum: Office Workstations in the Home, National Academy of Sciences, Board on Telecommunications and Computer Applications, Washington, DC.*

Mills, C. W. (1951). *White collar.* New York: Oxford University.

Mintzberg, H. (1973). *The nature of managerial work.* New York: Harper & Row.

Mintzberg, H. (1979). *The structuring of organizations. A synthesis of the research.* Englewood Cliffs, NJ: Prentice-Hall.

Miss Bills celebrates twenty-fifth Aetnaversary. (1950, June). *The Life Aetna-izer, 27*, pp. 16–17.

Moran, T. (1981). The command language grammar: A representation for the user interface of interactive computer systems. *International Journal of Man-Machine Studies, 15*, 3–50.

Moran, T. (Ed.). (1982). *Eight short papers in user psychology* (Report CIS-17). Cognitive and Instructional Sciences Series, Xerox Palo Alto Research Centers.

Mowshowitz, A. (Ed.). (1980). *Human choice and computers, 2.* Amsterdam, The Netherlands: North-Holland.

Mumford, E. (1982). *Designing secretaries.* Manchester, England: University of Manchester Business School Press.

Mumford, E. (1983). *Designing human systems for new technology. The ETHICS method.* Manchester, England: Manchester Business School.

Murphree, M. C. (1984). Brave new office: The changing world of the legal secretary. In K. Sacks & D. Remy (Eds.), *My troubles are going to have trouble with me: Everyday trials and triumphs of women workers* (pp. 140–159). New Brunswick, NJ: Rutgers University Press.

National Academy of Sciences. (1985). *Office workstations in the home.* Washington, DC: National Academy Press.

National Research Council. (1983). *Video displays, work, and vision.* Washington, DC: National Academy Press.

Nilles, J. M., Carlson, F. R., Gray, P., & Hanneman, G. (1976). *The telecommunications-transportation tradeoffs: Options for tomorrow.* New York: Wiley.

Noble, D. F. (1977). *America by design: Science technology and the rise of corporate capitalism.* New York: Knopf.

Nora, S., & Minc, A. (1980). Computerizing society. *Society, 17*, 25–30.

Noyelle, T. (1984, October). *Employment discrimination in the service economy.* Paper presented at the Conference on Women, Clerical Work, and Office Automation: Issues for Research, Organized by the Women's Bureau, U.S. Department of Labor and the Panel of Technology and Women's Employment, National Research Council, Washington, DC.

Nowotny, H. (1982). The information society. Its impact on the home, local community and marginal groups. In N. Bjørn-Andersen, M. Earl, O. Holst, & E. Mumford, (Eds). *Information society—for richer, for poorer* (pp. 127–145). Amsterdam, The Netherlands: North-Holland.

Nussbaum, K., & Gregory, J. (1980). *Race against time: Automation of the office.* Cleveland: Working Women Education Fund.

Offshore offices: Corporate paperwork sneaks out of the country. (1985, September). *Working Women*, 55.

Olson, M. H. (1971). *The logic of collective action.* Public goods and the theory of groups. Cambridge, MA: Harvard University Press.

Olson, M. H. (1982). New information technology and organizational culture. *MIS Quarterly, 6*, 71–92.

Olson, M. H. (1983). Remote office work: Changing work patterns in space and time. *Communications of the ACM, 26*, 182–187.

Olson, M. H., & Primps, S. B. (1984). Working at home with computers: Work and nonwork issues. *Journal of Social Issues, 40*, 97–112.

Onions, C. T. (1955). *The Oxford universal dictionary: On historical principles.* London: Oxford University Press.

Otway, H. J., & Peltu, M. (1984). *Managerial challenge of new office technology.* London: Butterworths.

Panko, R. (1984). Office work. *Office: Technology and People, 2*, 205–238.

Panko, R. (1982). Serving managers and professionals. *AFIPS Office Automation Conference Proceedings* (pp. 97–103).

Pape, A. (1983). The office of the future: Some sociological perspectives on office work and office technology. In E. Fossum (Ed.), *Computerization of working life* (pp. 125–144). West Sussex, England: Ellis Horwood Limited.

Pava, C. H. P. (1983). *Managing new office technology. An organizational strategy.* New York: Free Press.

Peters, T. J., & Waterman, R. H. (1982). *In search of excellence.* New York: Harper & Row.

Phelps, N. (1985). Mountain Bell: Program for managers. In *Office workstations in the home.* Washington, DC: National Academy Press.

Pierce, Robert H. (1951a). *Aetna Life in the 'twenties, 1923–1929: The sleeping giant awakes.* (History of the Aetna Life Insurance Company, monograph no. 23). Hartford: Aetna Life Insurance Company.

Pierce, Robert H. (1951b). *Home office staff, 1879–1922: A big family working.* (History of the Aetna Life Insurance Company, monograph no. 21). Hartford: Aetna Life Insurance Company.

Pierce, Robert H. (1951c). *Investments in depression, 1924–1941: Difficulties are numerous.* (History of the Aetna Life Insurance Company, monograph no. 26). Hartford: Aetna Life Insurance Company.

Pierce, Robert H. (1951d). *It makes me proud.* (History of the Aetna Life Insurance Company, monograph no. 30). Hartford: Aetna Life Insurance Company.

Podgursky, M. (1984, July). Sources of secular increases in the unemployment rate, 1962–82. *Monthly Labor Review, 107,* 19–25.

Pool, I. (Ed.). (1977). *The social impact of the telephone.* Cambridge, MA: MIT Press.

Pool, I. (1983). *Forecasting the telephone: A retrospective technology assessment of the telephone.* Norwood, NJ: Ablex.

Pool, I., Decker, C., Dizard, S., Israel, K., Rubin, P., & Weinstein, B. (1977). Foresight and hindsight: The case of the telephone. In I. Pool, (Ed.), *The social impact of the telephone* (pp. 1–32). Cambridge, MA: MIT Press.

Poppel, H. (1982). Who needs the office of the future? *Harvard Business Review, 60,* 146–155.

Porat, M. (1977). *The information economy: Definition and measurement.* (U.S. Department of Commerce, Office of Telecommunications, U.S. Government Printing Office Stock No. 003-000-00512-7). Washington, DC: U.S. Government Printing Office.

Porcella, M. (1985). *Controlling office automation: The case of computer-based text processing.* Unpublished master's thesis, University of California, Irvine.

Poza, E., & Markus, M. L. (1980). Success story: The team approach to work restructuring. *Organizational Dynamics, 8,* 2–25.

Quinn, R., & Shepard, L. (1974). *The quality of employment survey.* Ann Arbor, MI: Survey Research Center, Institute for Social Research, University of Michigan.

Reskin, B. (1985) *Women's work, men's work: Sex segregation on the job.* Washington, DC: National Academy Press.

Rittenhouse, R. (1985). *The social dynamics of computer-based text processing.* Unpublished doctoral dissertation, University of California, Irvine.

Rittenhouse, R., & Kling, R. (1984a). *Organizational choices in computer-based text processing* (Working paper). Irvine: University of California, Public Policy Research Organization.

Rittenhouse, R., & Kling, R. (1984b). *The social dynamics of computer-based text processing* (Working paper). Irvine: University of California, Public Policy Research Organization.

Ritzer, G. (1977). *Working: Conflict and change,* Englewood Cliffs, NJ: Prentice-Hall.

Robinson, A. L. (1977). Impact of electronics on employment: Productivity and displacement effects. *Science, 195,* 1179–1184.

Rogers, E. M. (1983). *The diffusion of innovations.* New York: Free Press.

Rogers, T., & Friedman, N. (1980). *Printers face automation.* Lexington, MA: Lexington Books.

Rosenberg, J. (1982). A psycholinguistic model for command names (Report No. CIS-20). Palo

Alto, CA: Xerox Palo Alto Research Center.

Rothberg, H. J. (1969). A study of the impact of office automation in the IRS. *Monthly Labor Review, 92*, 26–30.

Rumberger, R., & Levin, H. (1984, February). *Forecasting the impact of new technologies on the future job market* (Project Report No. 84-A4). Stanford: Institute for Research on Educational Finance and Governance, Stanford University.

Sadler, P. (1981). Welcome back to the "automation" debate. In T. Forester (Ed.), *The microelectronics revolution* (pp. 290–296). Cambridge, MA: MIT University Press.

Sapir, E. (1949). Communication. In D. Mandelbaum (Ed.), *Selected writings of Edward Sapir* (pp. 104–109). Berkeley, CA: University of California Press.

Schiff, F. W. (1983). Flexiplace: An idea whose time has come. *IEEE Transactions on Engineering Management, EM-30,* 26–30.

Schneider, L., Howard, R., & Emspak, F. (1985). *Office automation in a manufacturing setting: A case study. (OTA Contract No. 4330055).* Washington, DC: U.S. Office of Technology Assessment (OTA Contract No. 4330055).

Scott, J. (1982). The mechanization of women's work. *Scientific American, 247,* 166–187.

Shepard, J. M. (1971). *Automation and alienation: A study of office and factory workers.* Cambridge, MA: MIT Press.

Short, J., Williams, E., & Christie, B. (1976). *The social psychology of telecommunications.* New York: Wiley.

Shortliffe, E. (1976). *Computer-based medical consultations: MYCIN.* New York: Elsevier, Inc.: New York.

Silvestri, G. T., Lukasiewicz, J. M., & Einstein, M. E. (1983, November). Occupational employment projections through 1995. *Monthly Labor Review, 106,* 37–48.

Slimmon, J. B. (1925). [Memorandum to President M. B. Brainard, December 18]. Unpublished manuscript, Aetna Life Insurance Company, Circular Letters, Series C, Vol. 7, unnumbered page, Aetna Life and Casualty Company archives.

Slimmon, J. B. (1931a). [Bulletin to all officers and divisions heads, April 25]. Unpublished manuscript, Aetna Life Insurance Company, Circular Letters, Series C, Vol. 7, unnumbered page, Aetna Life and Casualty Company archives.

Slimmon, J. B. (1931b). [Bulletin to all officers and division heads, May 13]. Unpublished manuscript, Aetna Life Insurance Company, Circular Letters, Series C, Vol. 7, page 136, Aetna Life and Casualty Company archives.

Smith, D. (1984, June). *Future technologies and their impacts.* Paper presented at Bell Communications Research conference on Technology and the Transformation of White Collar Work, New Brunswick, NJ.

Smith, M. J. (1984). Human factors issues in VDT use: Environmental and workstation design considerations. *IEEE Computer Graphics and Applications, 4,* 56–63.

Soloway, E., Rubin, E., Woolf, B., & Bonar, J. (1983). MENO-II: An A-I programming tutor. *Journal of Computer-Based Instruction, 10,* 20–34.

Spinrad, R. J. (1982). Office automation. *Science, 215,* 808–813.

Strassman, P. A. (1980). The office of the future: Information management for the new age. *Technology Review, 82,* 54–65.

Strassman, P. A. (1985). *Information payoff: The transformation of work in the electronic age.* New York: Free Press.

Strom, S. H. (1985, February–March). *Technology, the office, and the changing sexual division of labor, 1910–1940.* Paper prepared for the Panel on Technology and Women's Employment Workshop, National Academy of Sciences: Washington, DC.

Sullivan, C. S., & Davis, L. E. (1980). A labour management contract and quality of working life. *Journal of Occupational Behavior, 1,* 29–41.

Tapscott, D. (1982). *Office automation: A user-driven method.* New York: Plenum Press.

Taylor, F. W. (1967). *The principles of scientific management.* New York: Norton. (Originally published in 1911).

Taylor, J. C. (1977). Job satisfaction and quality of working life: A reassessment. *Journal of Occupational Psychology, 50,* 243–252.

Taylor, J. C. (1978). The socio-technical approach to work design. In K. Legge & E. Mumford (Eds.), *Designing organizations for satisfaction and efficiency* (pp. 95–107). Westmead, England: Gower Press.

Taylor, J. C. (1979). Job design criteria twenty years later. In L. E. Davis & J. C. Taylor (Eds.), *Design of jobs* (2nd ed., pp. 54–63). Santa Monica: Goodyear.

Taylor, J. C. (1980). A day in the life of a word processing operator. *Words, 9,* 12–16.

Taylor, J. C. (1981). *Employee participation in the socio-technical analysis of a computer operations organization.* Paper presented at Quality of Working Life and the 80's Conference at Toronto, Canada. Summarized in Kolodny, H. & van Beinum, H. (Eds.), (1983). *The quality of working and the 1980s* (pp. 162–163). New York: Praeger.

Taylor, J. C., Gustavson, P. W., & Carter, W. S. (1985). Socio-technical systems and the adaptation of computer-aided design processes. In D. D. Davis (Ed.), *Dissemination and implementation of advanced manufacturing processes* (pp. 135–174). Proceedings of the Fourth Industrial/Organizational Psychology Scientist-Practitioner Conference. (National Science Foundation Report No. 18314184) Norfolk, VA: Old Dominion University.

Teger, S. L. (1983). Factors impacting the evolution of office automation. *Proceedings of the IEEE, 71,* 503–511.

Thoite, P. (1983). Multiple identities and psychological well-being: A reformulation and test of the social isolation hypothesis. *American Sociological Review, 48,* 174–187.

Toffler, A. (1970). *Future shock.* New York: Bantam.

Toffler, A. (1979). *The third wave.* New York: Morrow.

Tornatzky, L., & Klein, K. (1982). Innovation characteristics and innovation adoption-implementation: A meta-analysis of findings. *IEEE Transactions on Engineering Management, EM-29,* 28–45.

Tornatzky, L. G., Eveland, J. D., Boylan, M., Hetzner, W., Johnson, E., Roitman, D., & Schneider, J. (1983). *The process of technological innovation: Reviewing the literature.* Washington, DC: National Science Foundation.

Tornatzky, L. G., Fergus, E. O., Avellar, J. W., Fairweather, G. W., & Fleischer, M. (1980). *Innovation and social process.* New York: Pergamon Press.

Trist, E. (1981). The evolution of socio-technical systems: A conceptual framework and an action research program (Occasional Paper No. 2), Toronto: Ontario Quality of Working Life Center.

Turner, J. A. (1984). Computer-mediated work: The interplay between technology and jobs-claims representatives in the Social Security Administration. *Communications of the ACM, 27,* 1210–1217.

Uhlig, R. P., Farber, D. J., & Bair, J. H. (1979). *The office of the future: Computers and communications.* New York: North-Holland.

U.S. Bureau of the Census. (1970). Table 221: Detailed Occupation of the Experienced Labor Force and Employed Persons by Sex: 1970 and 1960. *Census of the Population, Characteristics of the Population, Vol. 1, Part 1.* Washington, DC: U.S. Government Printing Office.

U.S. Bureau of the Census. (1975). *Historical statistics of the United States, Colonial times to 1970.* (Parts 1 and 2). Washington, DC: U.S. Government Printing Office.

U.S. Bureau of the Census. (1980). *Census of population and housing, 1980: Public-use microdata sample A.* [MRDF] Washington, DC: Author.

U.S. Bureau of the Census. (1983). *Statistical abstract of the United States: 1984* (104th ed.). Washington, DC: U.S. Government Printing Office.

U.S. Bureau of Labor Statistics. (1961). *Industry wage survey.* (Bull. No. 1324: Average

Weekly Earnings, Selected Occupations). Washington, DC: U.S. Government Printing Office.
U.S. Bureau of Labor Statistics. (1965). *Impact of Office Automation in the Insurance Industry.* (Bull. No. 1468). Washington, DC: U.S. Government Printing Office.
U.S. Bureau of Labor Statistics. (1966). *Industry wage survey.* (Bull. No. 1569: Average Weekly Earnings, Selected Occupations), Washington, DC: U.S. Government Printing Office.
U.S. Bureau of Labor Statistics. (1971). *Industry wage survey.* (Bull. No. 1971: Average Weekly Earnings, Selected Occupations), Washington, DC: U.S. Government Printing Office.
U.S. Bureau of Labor Statistics. (1976). *Industry Wage Surveys: Banking and Life Insurance.* (Bulletin No. 1988, Table 67: Average Weekly Earnings, Selected Occupations). Washington, DC: U.S. Government Printing Office.
U.S. Bureau of Labor Statistics. (1978). *Employment and Earnings During 1978: An Analysis.* (Special Labor Force Report 218, Table 23). Washington, DC: U.S. Government Printing Office.
U.S. Dept. of Labor. (1961). *Industry wage survey.* (Bull. No. 1324: Average Weekly Earnings, Selected Occupations). Washington, DC: Bureau of Labor Statistics.
U.S. Dept. of Labor. (1965). *Impact of office automation in the insurance industry.* (Bull. No. 1468). Washington, DC: Bureau of Labor Statistics.
U.S. Dept. of Labor. (1976). *Industry wage surveys: Banking and life insurance.* (Bull. No. 1988, Table 67: Average Weekly Earnings, Selected Occupations). Washington, DC: Bureau of Labor Statistics.
U.S. Dept. of Labor. (1978). *Employment and earnings during 1978: An analysis* (Special Labor Force Report 218, Table 23). Washington, DC: Bureau of Labor Statistics.
U.S. National Commission on Technology, Automation, and Technological Change. (1966, February). *Technology and the American economy.* Washington, DC: U.S. Government Printing Office, Serial No. 0-788-561.
U.S. Personnel Classification Board. (1929). Salaries for routine clerical work in private industry–1929. Cited in: Life Office Management Association. (1932). *Clerical salaries in the life insurance business.* Fort Wayne, IN: Life Office Management Association.
Walton, R. E. (1975). The diffusion of new work structures: Explaining why success didn't take. *Organizational Dynamics, 3,* 2–22.
Walton, R. E. (1977). Successful strategies for diffusing work innovations. *Journal of Contemporary Business,* 1–21.
Walton, R. E. (1979) Work innovations in the United States. *Harvard Business Review, 57,* 88–98.
Weber, C. (1959). Impact of electronic data processing on clerical skills. *Personnel Administration, 22,* 21–26.
Whalen, T. (1983). Office technology and socio-economic change: 1870–1955. *IEEE Technology and Society Magazine, 2,* 12–18, 29.
Williams, L. K. (1981). Office automation—some problems and opportunities. *Office Automation Conference Digest* (pp. 261–264). Houston: AFIPS.
Wobbe-Ohlenburg, W. (1983, June). *Robots replace not only simple but also advanced tasks.* Goettingen, West Germany: Soziologisches Forschungsinstitute Goettingen (SOFI).
Wynn, E. (1982, April). *Linking user responses to the decision chain.* Paper presented at the Office Automation Conference, San Francisco.
Yin, R. K. (1978). *Changing urban bureaucracies: How new practices become routinized* (Report No. N-2277-NSF). Santa Monica, CA: The Rand Corp.
Yin, R. K., Quick, S., Bateman, R., & Marks, E. (1978). *Changing urban bureaucracies: How new practices become routinized—Appendixes.* Santa Monica, CA: Rand Corp.
Young, R. M. (1981). The machine inside the machine: Users' models of pocket calculators. *International Journal of Man-Machine Studies, 15,* 51–85.

Zuboff, S. (1982, September–October). New worlds of computer-mediated work. *Harvard Business Review, 60,* 142–152.

Zuboff, S. (1982, Winter). Problems of symbolic toil. *Dissent, 29,* 51–61.

Author Index

D

E

F

G

H

T

U

V

W

Y

Z

Subject Index

A

B

C

D

E

O

P

Q

R

S

U

V

W

porté lorsqu'elle subissait ses traitements de chimio, tu te dis avec fierté que tu aurais pu être une belle fille, faire tourner bien des têtes, et qu'au fond il s'en faut de peu pour être une femme ou un homme puisqu'on est toujours un peu des deux.

Il y a des choses qui nous arrivent et il y a des choses qu'on contrôle, c'est ce que tu as expliqué à ton chum quand tu es revenue de ton séjour en Italie où tu as eu une aventure avec un congressiste rencontré par hasard, et plutôt que de te laisser comme tu désirais secrètement qu'il le fasse, il t'a expliqué qu'il avait toujours confiance en toi et qu'il souhaitait voir si vous pouviez continuer, ce qui te désespère puisque, si tu as l'énergie de le tromper, tu ne trouves pas en toi la force de rompre, un peu par habitude, un peu par lâcheté, car tu sais que les crêpes qu'il te prépare tous les dimanches risquent de te manquer.

L'été de tes onze ans, alors que tu avais été sélectionné pour jouer dans les Aigles de Neufchâtel, on t'a appris que tu avais la mononucléose et que ta rate avait anormalement enflé, à un point tel que le médecin t'a interdit de jouer au soccer, et même si tu ne ressentais pas la fatigue typique de cette maladie, tu as dû te résigner à regarder les matchs depuis les estrades, où tu rêvais de remplacer un joueur in extremis, de te sacrifier et de mourir en pleine gloire sur le terrain des suites d'une hémorragie interne, juste après avoir marqué le but de la victoire, au lieu de quoi tu te tournais les pouces en regardant ton équipe perdre.